ATM Switches

ATM Switches

Edwin R. Coover

Artech House
Boston • London

Library of Congress Cataloging-in-Publication Data
Coover, Edwin R., 1942–
  ATM switches / Edwin R. Coover.
    p.   cm.
  Includes bibliographical references and index.
  ISBN 0-89006-783-X (alk. paper)
  1. Asynchronous transfer mode.   I. Title.
TK5105.35.C66   1997
004.6'6—dc21   97-12575
                                                      CIP

British Library Cataloguing in Publication Data
Coover, Edwin R.
  ATM switches
  1. Asynchronous transfer mode
  I. Title.
  621.3'8216

  ISBN 0-89006-783-X

Cover design by Darrell Judd

© 1997 ARTECH HOUSE, INC.
685 Canton Street
Norwood, MA 02062

International Standard Book Number: 0-89006-783-X
Library of Congress Catalog Card Number: 97-12575

10 9 8 7 6 5 4 3 2 1

Contents

Preface

From its inception, this volume has followed a twisted road. Originally conceived as a book on one of several promising new broadband technologies, the appearance of at least a dozen books on the emerging asynchronous transfer mode (ATM) standard clearly obviated that need. At the same time, the introduction of commercial ATM switches and components, and their peculiar interpretations and implementations of the standard, prompted a reappraisal. What is going on here? Has ATM, the much-touted "unification" technology, a single way of handling voice, video, and computer data across conventional local area network/metropolitan area networks/wide area network (LAN/MAN/WAN) boundaries, been highjacked?

It would seem so. First, a number of small startup companies such as FORE and Lightstream charged on the scene with "prestandard" LAN products. Their motto seems to be, "data can't wait!" They, in turn, have been followed by seemingly every company that produces time division multiplexers (TDMs), routers, LAN hubs, packet or frame relay switches, or central offices switches. The credo is that ATM is "The Next Big Thing."

The level of sheer hype on ATM makes even Microsoft's Windows 95 launch look modest. Does ATM present "Integration Technology AND End-to-End Technology?" Is the high-speed ATM network the "Backplane of the Metacomputer?" Will ATM result in the "Death of Distance?" Not very likely. Indeed, if anything, there is a bit of embarrassment on just where all this speed and elegant technology will be put to use. In the language of the industry flak, there is no acknowledged "killer app"—a technology-enabled breakthrough application. Instead, there is a host of contenders or perhaps pretenders: Lotus' (now IBM) Notes, desktop videoconferencing, distance learning, and various applications involving compressed video.

The ATM market is a gold rush, complete with Chaplinesqe aspects, but mostly it is messy and wild. It is the most exciting technology since the Ethernet. Were it not for the Ethernet precedent, one would be tempted to observe that there are too many companies for anyone to make any money; indeed, it may be just the beginning. Nonetheless, like most gold rushes, many will return (relatively) empty-handed. Lightstream was bought by router Goliath Cisco Systems, who subsequently gobbled up StrataCom. BBN's ATM switch became Agile Networks' ATMizer. Startup Centillion Networks was acquired by Bay Networks, itself a merger of Wellfleet and SynOptics. And just before this book went to press, Cascade was acquired by Ascend, a fast-rising vendor of remote-access products.

From the commercial product point of view—the view of this work—it is the exciting story of when a promising technology meets the magic of the marketplace, and the unexpected happens.

Acknowledgments

I have sometimes seen writing a book as a centripetal act in a centrifugal universe. Over the four years I have been researching and writing this one, the centripetal act has been much aided by the comments of colleagues (or ex-colleagues) at the MITRE Corporation. Rick Tibbs, Rob Rice, Randy Mitchell, David Koester, Satish Chandra, Susmit Patel, Rich Verjinski, Doug Bamford, and Tom Ramsay have influenced me on a number of communications issues. Janet Park produced the figures and tables, and they reflect her considerable skill and care. The staff at the MITRE library, including some now departed, provided the resources and an environment that enabled sustained investigation. They included Sherri Lieberman, Vi Jefferson, Hugh Keleher, John Huddy, Win Kutz, Shelia Chatterjee, Jean Watson, and Belinda Parker. Mark Walsh, acquisitions editor at Artech House, has been both supportive and patient. Artech staff Lisa Tomasello, Pamela Ahl, Beverly Cutter, Matt Gambino, Kimberly Collignon, Julie Lovelock, Roberta de Avila, Alexia Rosoff, Darrell Judd, Kate Feininger, and Judi Stone provided timely advice on a large span of issues and, in the process, averted some real howlers. Finally, I owe a considerable debt to an unknown outside reader who not only caught a number of embarrassing errors, but whose witty marginal comments backed me off from a number of judgments that appeared, in retrospect, excessively ex cathedra.

Acknowledgments

Introduction

1

1.1 WHY ATM?

One can identify three developments of the 1980s that impelled the development of high-speed switching.

The first was the continuing mismatch between computer channel speeds and the public transmission network. Twenty years ago it was "MIPS talking to Kbps;" now, it is close to "GIPS talking to Mbps." Although the transmission networks could, by means of several tiers of multiplexing and demultiplexing, partially accommodate this torrent of zeros and ones, it was neither economical nor did it make sense intuitively to multiplex and demultiplex bit streams whose end use was a high-bandwidth device. To provide an example, a Cray supercomputer with a 100-Mbyte/s channel (800 Mbps) driving a graphics application would have had to be muxed into multiple DS-3 (45 Mbps) streams and reassembled. The costs of this sort of mismatch simply ensured that for such applications, the cheapest route was usually to move the application to the supercomputer or the supercomputer to the application.

The second development was the widespread use of graphical representation. Widely used as a means of visualization and/or data reduction, the bit streams (even when compressed) again starkly outlined the mismatch between many graphical applications and an economic means to remotely distribute them. For example, a megapixel display (800 × 1200) using just 16-bit color graphics and employing only 15 updates per second generates a 230.4-Mbps bit stream. Although this 230.4-Mbps bit stream may (depending on the application's requirements) be satisfactorily compressible onto a DS-1 (1.544 Mbps) circuit, a leased DS-1 facility was often unjustifiable for a single high-bandwidth application; further, the obvious lower cost alternative, dial-up DS-1, is not ubiquitously available.

The third development was in response to the first two and consisted of proposals by standards groups for switching schemes that would economically accommodate a variety of types of high-speed bit streams on a shared media. The American National Standards Association (ANSI) considerably extended its local area network (LAN)-oriented fiber distributed data interface (FDDI) to FDDI II, making it a contender for LAN applications requiring both high-speed and isochronous (regularly timed) flows. Another ANSI proposal for the local area resulted in the fiber channel specification, a high-speed multimedia LAN running from 266 Mbps to 1 Gbps. From the Institute for Electrical and Electronic Engineers (IEEE) came the dual queue distributed bus (DQDB) technology, originally intended as a metropolitan area network (MAN) standard. Last, and the focus of this volume, the Consultative Committee for International Telephone and Telegraph (CCITT) proposed the asynchronous transfer mode (ATM) standard for wide area networks (WANs). The initial proposals for ATM assumed use of the synchronous optical network (SONET) transmission system in the United States or synchronous digital hierarchy (SDH) in Europe, but the earliest uses of ATM have been on non-SONET LANs. As all the above technologies overlap, and widespread use is more dependent on the facility with which interfaces can be set in silicon (and the subsequent production volumes and costs), it was by no means clear initially which one, or even which two, would emerge as the most popular among the user communities.

By the mid 1990s ATM has emerged as the leading contender with a number of companies fielding products ranging from network interface cards (NICs) for workstations through LAN-to-ATM concentrators to ATM switches in various sizes. The attraction of the ATM standard rests on several features. First, by employing short, 53-byte "cell" technology (essentially, fixed-length fast packets) it could successfully mix a wide variety of traffic types (voice, video, fax, and varieties of computer data). Second, the comparatively straightforward cell technology aptly lent itself to hardware implementation, and henceforth to very high speeds (currently to 622 Mbps, but potentially into the low gigabits).

Readers evaluating the ATM should be also be aware of a larger issue—ATM's possible role as a unifying technology. Divorced from the grubby realities of installed bases and return on investment, ATM is a technology that can transfer huge volumes of voice, computer data, video, and facsimile traffic at very high speeds and combine them for transmission efficiency, yet separate them for delivery within variable quality of service (QoS) guidelines. Combine this promise with ATM's geographical versatility—the local area, the metropolitan area, the wide area—and it is easy to see why ATM is often hyped as a sort of technological second coming. This book views this technological second coming as highly unlikely, at least in the foreseeable future. Yet, one cannot ignore the promise of the potential benefits of a single, high-speed, multimedia standard:

- Greater transmission efficiencies;
- Easier maintenance, fewer boxes in the string (equipment and/or links to deliver a service);
- Eliminating protocol conversion;
- Eliminating speed conversion;
- Eliminating disparate clocks;
- Encouraging economies of scale.

The intent of this book is to survey the fruits of the "first generation" ATM technology—the first commercial products using ATM technology, their network role, and their purported capabilities. It also intends, along the way, to include modest amounts of benign propaganda: the need for broadband communications, the capabilities of ATM, and where ATM fits in the current telecommunications milieu.

1.2 THE STATE OF THE STANDARD

The CCITT completed the proposed the ATM standard for WANs in 1988. Relative to the big picture, ATM forms but one portion of the broadband-integrated services distributed network (B-ISDN) standard; the core standards are schematized in Figure 1.1.

Since 1988, the ATM has been progressively defined by several groups. Most conspicuous has been the work of a large number of industry-dominated committees working under the umbrella of the ATM Forum. They started out clarifying the use of the standards developed by the International Telecommunications Union-Telecommunications (ITU-T, successor to the CCITT) and the ANSI T1S1 committee and ended up creating whole new specifications such as the User-to-Network Interface (UNI) standard version 3.1 and the LAN Emulation (LANE) Phase 1 protocols. Today, the ATM Forum can be seen as the de facto source of standards for private ATM networks. (The ongoing fruits of their labors can be viewed on the World Wide Web (http://www.atmforum.com/).

On the public network side, another influential group has been the Internet Engineering Task Force (IETF). It has focused on adapting the Internet protocol suite (IPS) to ATM and has been responsible for Request for Comment (RFC) 1483 (Multiprotocol Encapsulation), RFC 1577 (Classical IP over ATM Protocol), RFC 1755 (Signaling Guidelines for Classical IP), and others.

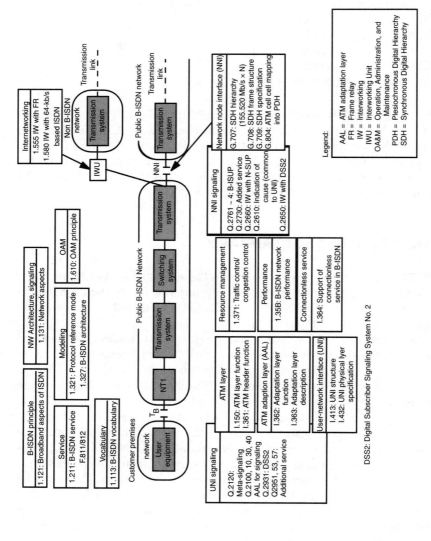

Figure 1.1 The broadband-integrated services distributed network standard (*Source:* Asatani, K., "Standardization of Network Technologies and Services," *IEEE Communications Magazine,* July 1994, p. 87. © 1994 IEEE.)

At least potentially, another significant player could be the Microsoft Corporation. It is working on a applications programming interface (API) for ATM called "winsock." Empowering the user to request, if not actually specify, a QoS, would be an enormous step toward the long-sought goal of "bandwidth on demand." Were Microsoft to incorporate the winsock API in future editions of its popular operating systems, it would be a powerful force in popularizing ATM at the desktop level.

As of this writing (Summer 1996), ATM standards are complete to the point that something over 40 switching products have appeared from over 30 firms. Nonetheless, a number of important areas (congestion control, for one) have been agreed upon so recently that they have not yet appeared in commercial products. Other areas remain undefined, and it is clear that the ATM Forum will continue to add important elements well beyond 1996. (Indeed, if ATM is successful and finds many uses, the work of the ATM Forum will become perpetual [1].)

From the perspective of a communications planner, when is it likely to see the combination of a full set of ATM service capabilities for voice, video, and computer data along with industry-wide switch interoperability? If one excludes interoperable usage-based accounting and network management diagnostics (software-intensive and likely the last to appear), it is probably another two or three years distant. If one includes interoperable accounting and network management features, four years is more likely. Nonetheless, the distinction is worth noting. Most of the ATM switches sold today are designed to provide high-speed data services in the local area where accounting is not required and network management among homogeneous switches can be rudimentary. Thus, when the question is posed as to when ATM will be genuinely of use, the answer depends on the application. The correct answer can be today.

1.3 ATM CONSTITUENCIES TODAY

It is less the case that today's networks cannot handle large data volumes but that they cannot handle them economically. Today's carrier networks were historically sized to carry voice traffic in 64-Kbps units and various common aggregates (24 voice circuits in DS-1, 672 in DS-3). The central requirement of broadband services is the ability to handle much larger units—largely video traffic, most likely between 6 to 60 Mbps per channel—at comparably low costs.

Even for video transport, ATM is not a sure thing. ATM switching will only be used in video delivery if the commercial-off-the-shelf (COTS) products are cheap and efficient enough to offset that fact that video would be more efficiently moved in larger hunks than 48 octets. Hybrid or superset products are likely, such as IBM's Planet switch that moves data in larger hunks for efficiency but recognizes ATM cells as well and thus function as an "ATM switch." Essential to ATM's primacy as a video mover is a timely and efficient variable bit rate (VBR) standard coupled with the availability of cheap COTS components.

What are these video services and who is to pay for them? Overwhelmingly, they will be entertainment—movies viewed at home or while traveling, high-definition television (news, sports, shopping), and interactive games (including gambling). This list may strike communications visionaries as a rather banal use for elegant technology, but, at least for now, only these areas offer a proven record of popularity and revenue. More importantly, the financial success of these sorts of commercial services will indirectly enable a second tier of services. This second tier will be business-oriented, such as the economical transport and interchange of a large variety of video and graphics products (meetings, medical and weather data, industrial design, etc.).

Another shortfall for the visionaries will be the mode of introduction of ATM-based broadband services. In theory, ATM's multimedia capabilities could allow a number of extant networks (public or private) handling voice, data, or video to be "rolled up" into one big network, offering superior efficiency. But that approach, where all or a significant portion of a switching infrastructure is replaced, is highly unlikely. First, there is the installed base of current equipment. For instance, there are in the neighborhood of 20,000 large switches in the United States' public switched telephone network (PSTN) alone. Any sort of swap-out would have to offer immense financial benefits—not the sort of incremental, logical-multiplex gain promised by ATM's cell technology. Second, the current marketing direction is to more network "intelligence," more software-defined services and provisioning flexibility than size for size's sake. Already the larger central offices and major fiber highways threaten the reliability of PSTN. As numerous critics have remarked, the pursuit of greater operating economies by both local and long-distance carriers has already resulted in too many eggs in too few baskets. With disarming frequency, "fortress" style central offices fall victim to fire and flood; fiber cuts disrupt the huge bandwidth bundles, leaving vendors unable to divert traffic without outages; and software bugs humble switching systems [2]. Greater resource aggregation, whether of links or nodes, is simply undesirable. Third, and most important, investors in video distribution schemes want the maximum return upon their investment, virtually guaranteeing that ATM-provided video services will be provided in an "overlay" or supplemental manner, with the most lucrative markets (hotels, casinos, cruise ships, international flights) addressed first.

Similarly, ATM-based broadband services will likely employ suboptimal, conservative, operating practices. Switch designers and engineers have spent enormous effort on how to best protect ATM switches from the anticipated, huge, multimedia-based traffic surges. There have been numerous proposals, some of them quite clever. Most are "policy-based"—that is, related to the class of service promised the customer— and include the quiescing of lower priority traffic, increased delays, dynamically altered coding schemes (i.e., substituting 32-Kbps ADPCM voice for 64-Kbps PCM voice), and outright discarding of low-priority data. Chances are, for all the effort expended, ATM customers will never see the fruits of these particular labors, and for several reasons. First, vendors do not like to sell overtly probabilistic services, regardless of rates. Up-front complexity increases the cost of the sale. Later, invariably, there is disappointment (or haggling) and it tends to taint the vendor and the service. Second, communications planners are conservative people and do not like to be embarrassed. User complaints about congestion or poor performance sting more than management criticisms on cost. Having been through it many times before, they are used to being lowballed on traffic estimates, especially on computer data requirements. High one-time charges for installation also reinforce the tendency to "pad up." And rarely is the environment so cost-sensitive that they are unable, somewhere in the process, to "round up" on the bandwidth. The net effect of these tendencies will be to subvert the various ATM's traffic-shedding schemes and maintain the status quo—the "overengineered" network that works as advertised but is considerably more expensive than it has to be.

Several observations concern the complexities of switch design. Switch designs are born in conflicting (and changing) customer requirements, the promise of cheap hardware (high profits), the necessity for expensive software (essential customer features, product delays, bugs), and a continual tug of war between marketing (that wants the product of the future now) and engineering (that wants the product to work as advertised).

Before moving to the technical challenges of high-speed switching, it is best to view switching and the problems encountered from a very great distance. One, for instance, can liken the switching of many users and many services to the handling of passengers, freight, and mail at a large airport. Passengers may arrive by air, taxi, rental car, limo, bus, private automobile, and, in some locales, subway or train. Freight and mail arrives by air or truck. All encounter lines and delays—freeway slowdowns, parking lots, bus or train pickups, ticket and baggage checks, security checks, and gate checks. All have to be information-processed and routed, typically several times—reservations, credit cards, passports, tickets, boarding passes, baggage receipts, and baggage routing tags. Where possible, parallel processing speeds throughput: multiple lanes, multiple lots, multiple carrier areas, multiple ticket windows, multiple conveyors, multiple security stations, multiple people movers, multiple gates, multiple air carriers, multiple aircraft, and multiple runways. Contrariwise, the airport is liable to systemic pathologies—rain, fog, ice, snow, extreme heat and cold, equipment malfunctions of every type at every stage—as well as traffic fluctuations based on time of day, day of week, known peaks (Thanksgiving, Christmas, Spring vacation) and unknown peaks (the only airport open). Airports and large switching systems, constitute a well-known challenge in information theory: moving highly variable items with a highly variable information content. Today's airports provide the juncture between one's Porsche (130 mph) and the Boeing 777 (400 kn) just as today's ATM switches link a Cray's channel (800 Mbyte/s) and the OC-3c (155 Mbps) leased fiber-optic facility. Those close to either system can attest to international flights where more time was spent on the ground than in the air, or to the proclivity with which computer and communications equipment takes unexpected and variable-length work breaks. Rocked between the worst and the best, we often find ourselves thankful for uneventful but slow progress.

1.4 SOURCES, SINKS, AND THE FUTURE

To use an agribusiness simile, a data source is like grain in the silo; a sink is like grain in a truck. As noted earlier, the likeliest sources for broadband networks will be the communications/entertainment conglomerates; the sinks, today's world consumers of video products. This may seem to some as overly intellectually economical, or, from a haute culture point of view, that of trash begetting trash. It is not, or not necessarily. Despite H. L. Mencken's taunt "that no one ever went broke underestimating the taste of the American people," in fact people have. Rock 'n roller Bruce Springsteen has already lamented the phenomenon in a song titled "59 Channels and Nothing's On." What one will see, when compared with today, is a torrent of bits, or choice if one likes to think of it thus. The nature of the choices reflect the proclivities and tastes of a mostly wired world of near-universal capitalism and cultural (if not political) democracy. Three billion of the world's 5.6 billion people were estimated to have watched the 1994 World Cup final—far more than subscribe to any national or religious affiliation.

End Notes

[1] Marks, D. R., "A Forum's Work Is Never Done," *Data Communications*, April 1995, p. 130.

[2] McDonald, J. C., "Public Network Integrity—Avoiding a Crisis in Trust," *IEEE Journal on Selected Areas in Communications*, Vol. 12, No. 1, Jan. 1994, pp. 5–12.

The ATM Standard

2

Again, why ATM? The attraction of the ATM stems from a number of features in concert. First, by employing short, 53-byte "cell" technology (essentially, fixed-length fast packets), ATM can successfully mix a wide variety of traffic types (voice, video, fax, computer data) while maintaining strict quality of service (QoS) guarantees. Second, the comparatively straightforward cell technology aptly lends itself to hardware implementation, and henceforth to very high speeds (currently to 622 Mbps, but potentially into the gigabits). Third, ATM switches allow one (again within strict QoS strictures) to statistically multiplex the merged data streams to maximally use available bandwidth. Lastly, ATM has the potential to become a "unification" technology, a single way of handling voice, video, and computer data across conventional LAN/MAN/WAN boundaries. Potentially, at least, there are huge potential benefits associated with such a single, high-speed multimedia standard.

An overview of the ATM standard will be presented in the following pages. It will be brief for several reasons. Most people reading this book are presumed to have a general knowledge of the ATM standard, and, if not, there are a number books devoted to the subject. (An annotated bibliography is included in this volume.) Also, the standard is not really complete at this writing, and the ATM Forum will continue to produce significant additions beyond 1996. Finally, as already noted, most current ATM products have chosen to incorporate only certain ATM services and features, even in areas where the standard is well-defined. Many of the current implementors have viewed certain ATM capabilities as "dead on arrival." Because they may be wrong, and a subsequent generation of products may use these capabilities, it is worth reviewing the initial conception.

2.1 CELLS

A diagram of ATM's now-famous cell format appears in Figure 2.1. (Wits have remarked that all briefings involving ATM must be 53 slides long.) Note that with 5 octets of header and 48 octets of payload, the "going-in" overhead, *before* all the other overheads are piled on, is 9%.

The overtly political nature of this technological compromise on packet length (the voice constituency wanting them short and of a fixed length, the data constituency wanting them long and of variable length) remains bothersome to some. One study, however, tried to quantify just how inappropriate it was for data by trapping several weeks' worth of real TCP/IP traffic, simulating its transmission by ATM, and attempting to calculate what would have been the optimal cell size [1]. They came to two conclusions. First, their real TCP/IP traffic seemed to follow a "70% rule" whereby 70% of the total octets transferred were long packets, but that 70% of the total messages were short packets. Second, their conclusion on the optical cell size, based on their sample, was that the optimum cell size would have been around 60 to 80 octets—certainly in the ATM ballpark. One of the funniest commentaries compared the ATM 53-byte cell to the human nose, observing that, yes, observers have long pondered a number of design changes to improve it, but, on the whole, it works and its problems (e.g., congestion, leakage) are amenable to low-technology solutions (e.g., handkerchiefs) [2].

2.2 A LAYERED PROTOCOL

By the familiar ISO standards—which ATM does not follow—ATM would be characterized as a "layer 2/3 protocol." It assumes a synchronous physical layer (ISO 1) and accommodates copious physical framing and electrical options (T1, SONET, etc.) below it. Next step up, a data link (ISO 2) "sublayer" supports a huge number of logical and blocking options (SMDS, TAXI, etc.). At the data link layer proper is the distinctive

Cell (53 Bytes)					
Header					Information
Generic Flow Control	VCI/VPI Field	Payload Type Indicator	Cell Loss Priority	Header Checksum	Payload
4 bits	24 bits	2 bits	2 bits	8 bits	48 bytes

Figure 2.1 ATM cell structure. (*Source:* Vetter, R.J., "ATM Concepts, Architectures, and Protocols," *Communications of the ACM,* Vol. 38, No. 2, Feb. 1995, p. 34.)

53-octet envelope. Minimal error checking is supported at this level—a checksum on the cell header. The header contains addresses enabling a route hierarchy (*x* channels over *y* paths) and a code indicating the type of payload. At the network layer (ISO 3) there is a source address, a destination address (which may be a multicast group), and one of two kinds of address formats, depending on the application/network context.

If this appears rather straightforward, well it is not. As shown in Figure 2.2, in a sublayer above the ATM addressing is the ATM adaptation layer, whereby ATM deals with the real world of legacy systems (see Figure 2.1). Rather than a neat ISO 1/2/3 sandwich, it more closely resembles a Big Mac ("two all-meat patties"), with, for example, an Ethernet/IP layer 2/3 sitting on top of the ATM layer 2/3, with the ATM adaptation layer managing the necessary conversions.

2.3 CLASSES OF SERVICE

The ability to guarantee class of service to different applications is why the more rhapsodic of ATM commentators refer it as the "unification technology." More prosaically, it means that with decent planning, users with very different application needs (voice, video, several kinds of computer data) can use the same network links and be satisfied with the service, and that maybe their organization will even save money.

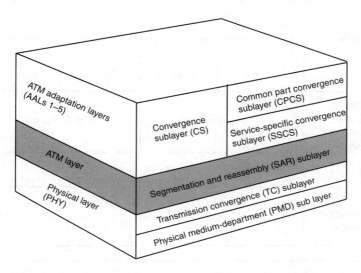

Figure 2.2 ATM layers and sublayers. (*Source:* Marks, D. R., "ATM from A to Z: A Definitive Glossary for Enterprise Network Managers," *Data Communications*, Dec. 1994, p. 113. © 1994 The McGraw-Hill Companies. All rights reserved.)

Figure 2.3 shows the service classes for broadband-integrated services networks (B-ISDN).

More recently, the ATM Forum's user-to-network interface (UNI)
has, for all practical purposes, redefined the B-ISDN service classes (pictured in Figure 2.3) into the following:

- *Continuous bit rate* (CBR), also called circuit emulation, which is
  intended for use with voice or uncompressed video.
- *Variable bit rate* (VBR), which is intended for compressed voice and
  compressed video but the service standards are still incomplete. Future VBR standards will come in at least two varieties:

  VBR-real time (RT), when there is a fixed timing relation between
  data samples, such as with variable bit rate video compression;

  VBR-nonreal time (NRT), when there is no timing relation between samples but a guaranteed QoS is required, such as when a
  frame relay connection with a committed information rate (CIR) is
  passed across an ATM network.

- *Available bit rate* (ABR), which is an untimed, variable data rate,
  "best effort" network service; feedback (flow control mechanisms)
  are employed to increase the user's allowed cell rate (ACR) and
  minimize cell loss. ABR is designed to handle bursty LAN protocols

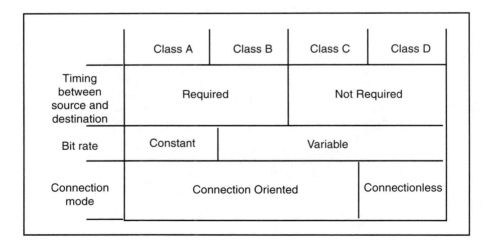

	Class A	Class B	Class C	Class D
Timing between source and destination	Required		Not Required	
Bit rate	Constant	Variable		
Connection mode	Connection Oriented			Connectionless

Figure 2.3 Service classes in B-ISDN. (*Source:* Vetter, R.J., "ATM Concepts, Architectures, and Protocols," *Communications of the ACM*, Vol. 38, No. 2, Feb.
1995, p. 34.)

and will be used for LAN emulation (LANE). The ABR service will, optionally, provide a guaranteed minimum cell rate (MCR).

- *Unspecified bit rate* (UBR), which offers no QoS guarantees, is what is available currently for bursty data traffic. Today's vendor implementations of LANE (for Ethernets and token rings) and RFC 1577 (for IP) use it. Several vendors offer an "enhanced UBR" featuring early packet discard (EPD), which mitigates the effects of cell loss on packet transfer. As it features neither flow control mechanisms nor adequate buffering to minimize the effects of large bursts, it is strictly *caveat emptor.*

Commercial vendors have gone every which way with these classes. Some low-end data switches offer only a single class, such as an enhanced UBR. Others boast as many as 32 separate classes of service. In an attempt to reduce configuration complexity, vendors have concocted new QoS descriptors such as "IPX" or "SNA." It is too soon to evaluate the efforts of these "short hands" but at least potentially they could impede heterogeneous switch interoperability.

2.4 SIGNALING

ATM networks employ signaling sequences to establish and release connections that are similar to those used in other connection-oriented networks. It is anticipated that the eventual standard will be very similar to that used in today's increasingly popular integrated services digital network (ISDN) service (often called "narrowband ISDN.") The sequence will consist of a formal user request for network resources and will negotiate the connection types (switched virtual circuit (SVC), point to point, or multipoint), call endpoints (ATM addresses), service parameters (which AAL to use), and virtual path and virtual channel identifiers (VPI/VCI) number allocation. In sum, although not yet final, the B-ISDN ATM standard for the UNI connection sequence is expected closely follow the ITU-T Q.931 standard for narrowband ISDN.

2.5 CALLS AND CONNECTIONS

ATM is connection-oriented, which, like a telephone call, means that a session must be set up before data is transferred, and terminated when completed. Unlike connectionless networks like those based on TCP/IP, where packet headers require source and destination addresses and sequence numbers (for resequencing), ATM puts limited functionality in the cell header. A connection, called a virtual connection (VC), is identified by a number (identifier) that has only local significance per link in the virtual connection.

The ATM virtual connection is identified by two subfields of the cell header: the virtual channel identifier (VCI) and virtual path identifier (VPI). The VCI field identifies dynamically allocable connections; the VPI field identifies statically allocable connections. Virtual channels are characterized by a VCI assigned at call setup. As noted, a VCI has only local significance on a link between ATM nodes. When the connection is released, the VCI values on the involved links are released and it can be reused by other connections. Virtual paths are defined by the header VPI and support semipermanent connections between endpoints. This virtual path (or virtual network) concept enables the transport of large numbers of simultaneous connections while providing an efficient and simple management of network resources. Figure 2.4 shows how a single semipermanent virtual path can support numerous dynamically allocable virtual channels.

2.6 CONNECTION-ORIENTED

Once upon a time the call setup process was seen as a source of undesirable delay and part of the rationale for "connectionless" (packets routed dynamically routed at each intermediate node) data networks. With ATM switches, this call setup process is so rapid that it reopens the "connectionless" versus "connection-oriented" debate, and, more importantly, allows ATM to be a candidate technology for many more application types [3].

As a particular SVC may traverse a number of ATM switches, and each SVC must deliver call-specific QoS guarantees, a number of outcomes are possible with call setup. In the best case, when the call is made, a connection swiftly follows. Should the initial attempt fail at a particular node—an intermediary switch cannot guarantee the bandwidth and QoS

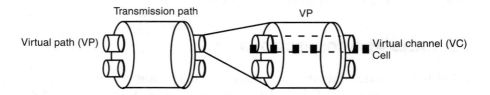

Figure 2.4 ATM network hierarchy. (*Source:* Saito, H., *Teletraffic Technologies in ATM Networks*, Norwood, MA: Artech House, 1994, p. 5.)

requested—then the connection attempt is "cranked back" to the last capable node and an alternate path attempted. In the worse case, such as a request for large bandwidth and stringent QoS guarantees in a heavily loaded network, an SVC request may be refused. Connection admission control (CAC) is portrayed in Figure 2.5.

The more serious issue with a connection-oriented service with ATM is accommodating legacy systems. Current telephone and video distribution systems are connection-oriented. Some data networks like X.25 and SNA are connection-oriented. But the very numerous IEEE 802.3/5/6 LANs that interconnect with TCP/IP and Novell's IPX are connectionless and have necessitated special support (LANE, RFC 1577).

2.7 TRAFFIC TYPES

ATM was designed from day one to support multiple media. Figure 2.6 shows how, from the perspective of the late 1980s, the media support needs requirements were viewed, and, following, how the AALs to accommodate them were structured.

2.8 ATM ADAPTATION LAYERS (AAL)

Sometime in the far future a wide range of computing and telecommunications may employ ATM in the "native mode." That is, network interface cards (NICs), switches of all sizes, transmission equipment, and perhaps even workstation buses and computer peripherals, will commu-

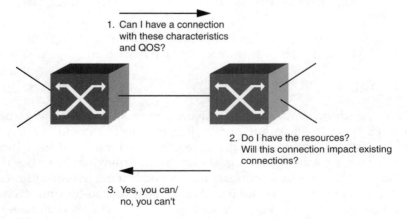

Figure 2.5 Connection admission control. (*Source:* Alles, A., *ATM Internetworking*, San Jose, CA: Cisco Systems, May 1995, p. 15.)

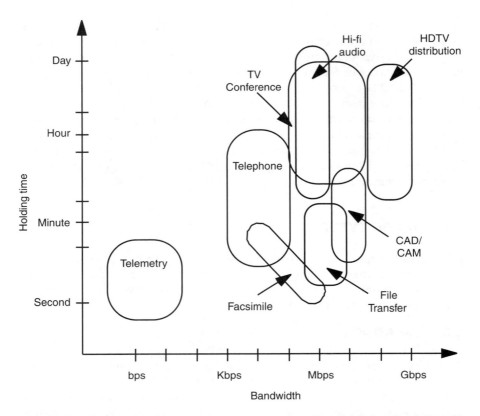

Figure 2.6 Traffic characteristics of various services. (*Source:* Saito, H., *Teletraffic Technologies in ATM Networks*, Norwood, MA: Artech House, 1994, p. 2.)

nicate in ATM cells. But for now, and probably for a very long time in the future, ATM-based communications will take place through AALs.

2.8.1   AAL 1

AAL 1 was designed for voice and uncompressed video and is often known as "circuit emulation" mode. It is similar to the old "nailed circuit" in that the bandwidth is dedicated for the duration of the connection and no gain is available from statistical multiplexing. Its critics have pointed out that as AAL imposes 1 octet of additional overhead for each 48-octet data cell, it is less efficient than present robbed-bit time division multiplexing. Defenders have pointed out its potential to eliminate certain multiplexing equipment such as digital access multiplexers (DACs) and (by enabling larger voice/data bundles) the ability to take advantage of quantity discounts in WAN bandwidth.

2.8.2 AAL 2

AAL 2 was designed for real-time traffic where the timing constraints were less rigorous than circuit emulation. Many believe it was designed for compressed video; economical video servers would be enabled by a combination of a comparatively low-volume, but variable, bit streams with preprocessing and postprocessing for compression and decompression, respectively. At this writing, Type 2 is unspecified by the ATM Forum and so matters of overhead and timing variation remain problematic. Critical reaction is schizophrenic. Some have pronounced it to be "dead on arrival." Others, looking ahead to the growth of video service in transportation and medicine, predict it will be the paramount application of the ATM technology.

2.8.3 AAL 3/4

AAL 3/4 traffic was intended to handle bursty traffic without cell loss. Designed for either bursty connection-oriented or variable rate connectionless traffic, it incorporates advanced features for sophisticated error checking (detection and correction of single bit errors, detection of multiple bit errors) and connection-level multiplexing. To accomplish these capabilities, it imposes a 4-octet overhead on every 48-byte data cell. Its data communications critics have charged that AAL 3/4 is simply overengineered and another 8% AAL overhead in addition to the basic 9% ATM overhead is outrageous. Its supporters see that argument as essentially religious in nature—in applications where cell loss is absolutely prohibited, at what layer (or layers) is one to do serious error checking? In the early going, most of the use of AAL 3/4 has been in ex-Bell Operating Companies (BOCs) providing high-speed switched multimegabit data service (SMDS) connections between the customer premises and frame relay (FR) switches. It is altogether possible that AAL 3/4 use and the commercial success of SMDS may be linked.

2.8.4 AAL 5

AAL 5 presents a second attempt to appease the data constituency. AAL 5 has been called the simple and efficient adaptation layer (SEAL), a pejorative reference to AAL 3/4. It does not allow connection multiplexing like AAL 3/4. It imposes no additional overhead on the 48-octet data portion but includes an 8-byte trailer in the last cells, which includes a 4-byte CRC. Ironically, AAL 5 provides better error detection than AAL 3/4 and is able to detect multiple bit errors and misordered cells. Just as most of the early use of ATM has been to provide high-speed data connections, most of these high-speed data connections have used AAL 5. Many low-end, LAN-oriented commercial products offer nothing but AAL 5.

2.9 ADDRESSING

With switched virtual circuits part of ATM's repertoire, there is the re-quirement for a standardized way to represent source and destination ad-dresses. (By contrast, with PVCs, connections and endpoints are defined and users need only to provide the network with a preallocated VCI/VPI.) However, for SVCs, the destination connection can change with each ses-sion and explicit addresses are needed. After the call has been mapped between UNIs, the VCI/VPI values are used to identify the traffic.

Similar to the compromise on cell length, ATM addressing resulted in another compromise between traditional voice and data constituen-cies. The ATM address is modeled on the OSI network service access point (NSAP), defined in ISO standard 8348, and ITU-T standard X.213, Annex A. There are four permitted formats, the first two of them "com-puter network style" and the latter two in the "telco style." A private net-work's UNI must support the first three; a public network's UNI must support either the first three or just the last. They are as follows:

1. Data Country Code (DCC) format;
2. International Code Designator (ICD) format;
3. ITU-T E.164 Private format;
4. ITU-T E.164 Public format.

Figure 2.7 shows the four formats and their constituent fields.

More than an agreement to disagree, this two-pronged approach to ATM addressing has several important benefits. First, the ATM Forum shunned the temptation to reinvent the wheel and used well-established formats—the ISO's NSAP for the computer-style addresses and the ITU-T's E.164 for the telco-style addresses. Second, the approaches are complimentary; it is very likely that large, ATM-based networks will eventually incorporate public switched connections for occasional links or extra redundancy. Third, for private ATM networks, the addressing is kept local (a registry is not required) and is similar to such well-known and widely used addressing conventions as IP. For example, in the pri-vate DCC format, the authority identified by the IDI is responsible for the

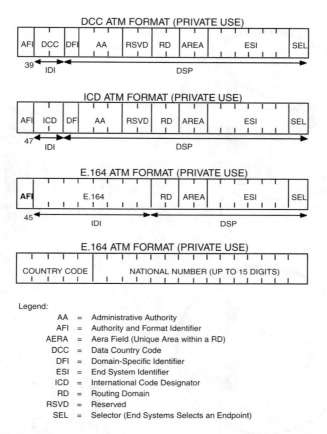

Figure 2.7 ATM address formats. (*Source:* Goralski, W. J., *Introduction to ATM Networking*, New York, NY: McGraw-Hill, 1995, p. 209. © 1995 The McGraw-Hill Companies. Reprinted with permission.)

contents of the RSVD, RD, and AREA fields. In the ICD version, the same latitude is extended to the AA, RSVD, RD, and AREA fields. In the private E.164, the RD and AREA fields are locally controlled.

2.10 NETWORK INTERFACES

Network interfaces are the protocols and conventions whereby one part of the network communicates with another. As the complexity (and attendant cost) varies, there are a number of types in use. For labs or local networks, these interfaces may be unimportant, but they are absolutely crucial for the operation of a large-scale network and are equally crucial

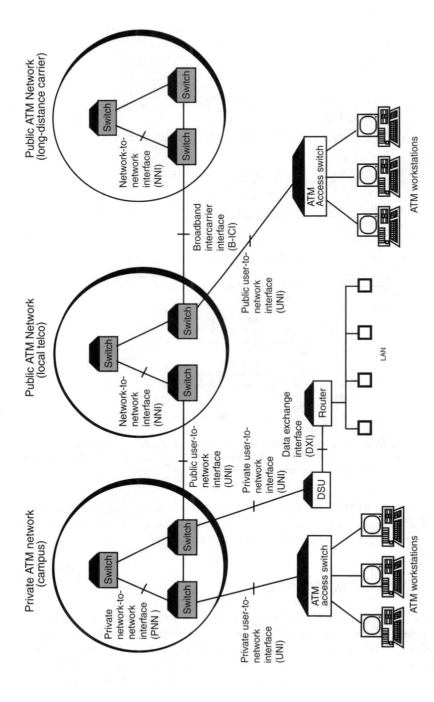

Figure 2.8 ATM logical network interfaces. (*Source:* Johnson, J. T., "The ATM Circus Gets a Ringmaster," *Data Communications,* March 21, 1993, p. 43. © 1993 The McGraw Hill Companies. All rights reserved.)

for heterogeneous switch interoperability. The ATM Forum has recently completed work on the following network interfaces:

- Data exchange interface (DXI);
- User-to-network interface (UNI);
- Network-to-network interface (NNI);
- Broadband intercarrier interface (B-ICI);
- Private network-to-network interface (PNNI).

Figure 2.8 shows the various contexts of the ATM logical network interfaces.

2.11 SUMMARY

This chapter has provided a minimalist overview to the ATM technology, with an emphasis on the original CCITT conception. At its most basic, ATM is connection-oriented, fast packet (cell) switching. It not only incorporates considerable technopolitical compromise but wholly incorporates proven technologies from the voice and data constituencies in addressing and signaling. The current ATM adaptation layers (AAL 1-5), UNI service classes (CBR/VBR/ABR/UBR), and network interfaces (DXI, UNI, NNI, B-ICI, PNNI) represent today's thinking on how to best employ ATM's cell technology. They are, however, by no means the last word, and should the ATM technology become popular, one will likely see new AALs, service classes, interfaces, and, as advertising folk say, "much much more."

End Notes

[1] Armitage, G. J., and K. M. Adams, "How Inefficient is IP Over ATM Anyway?," *IEEE Network*, Jan./Feb. 1995, pp. 18–26.
[2] Stevenson, D., "Electropolitical Correctness and High-Speed Networking, Or Why ATM Is Like A Nose," in Viniotis, Y., and R. O. Onvural, *Asynchronous Transfer Mode Networks*, New York, NY: Plenum Press, 1993.
[3] Bellman, J., (ed.), "ATM Practical Implementations Here and Now," Advertising Supplement to *Data Communications*, Feb. 1996, pp. A1–A12.

The Technical Challenges of High-Speed Switching I

3

ATM is not a bigger, brighter light bulb. Its benefits do not include simplicity. It will force one to do more planning and to understand one's application (and network) in greater depth than predecessor smart multiplexer or virtual private network solutions. However, first one must understand the basic problems endemic to high-speed switching and networks, the design compromises, and particular pathologies.

3.1 INTEROPERABILITY WITH LEGACY SYSTEMS

Legacy systems is a slightly pejorative term for the computer and communications systems one is currently using. The clear implication is that they await replacement by something better. Like money left one by a distant relative, there is also the implication that these systems were not one's direct doings. Less metaphorically, legacy systems were designed differently.

Unlike ATM, they generally used different technologies for the local area and wide area. WANs used leased circuits and SNA or DECnet or public networks (or public network technologies) and ran TCP/IP or X.25. LANs used 3270 or asynchronous terminal protocols or Ethernets or token rings running Novell's Netware or TCP/IP. Except for the significant exception of TCP/IP, the legacy of legacy systems was different technical approaches to the local and wide area and, inadvertently, extra complexity.

Another design difference is that of connection-oriented and connectionless. Current telephone and video distribution systems are connection-oriented. Some data networks like X.25 and SNA are connection-oriented. But the very numerous IEEE 802.3/5 LANs that interconnect with TCP/IP and Novell's Internetwork Packet Exchange (IPX) are connectionless and have necessitated a complex accommodation. At this writing it is too soon to evaluate whether this complex bridging of network types will be been successful.

Meanwhile, in the marketplace, where the time-honored way to market share and niche dominance is being the "first with the mostest," what does one do when a standard is clearly needed and yet the standard is not there? The answer is the "prestandard," or the proprietary implementation that makes the product useful to users and represents the company's best guess at what the standard implementation will look like. As a consequence, most of today's working ATM networks employ "prestandard" software. To their credit, realizing the importance of interoperability, few have tried to carve out proprietary niches; almost all have rushed to implement the new standards once blessed by the ATM Forum.

In the WAN area, ATM switches are likely to eventually replace smart multiplexers, packet switches, routers, and frame relay equipment. Depending on the context, ATM can provide more efficient multiplexing, better QoS guarantees, or simply speed. The anticipation is that the data communications protocols transplanted (SNA, DECnet, TCP/IP, IPX) will be encapsulated using AAL 5.

LANs can be viewed as a special case of legacy protocols. First, their ubiquity makes a smooth ATM transition an absolute necessity. Second, Ethernet, the most popular, is so unlike ATM that its efficient accommodation poses a significant technical challenge. Third, as noted earlier, providing higher performance LAN services has been the raison d'être of the early ATM switches.

With multitudes connected to Ethernet or token ring LANs, congestion at servers or network gateways often becomes severe. As a consequence, many organizations have moved to what are, in effect, two-level hierarchical LANs, often called "collapsed backbones." These feature Ethernet or token ring work group LANs, often with local servers, that are then attached to departmental LANs for communications between departments, whether local or distant. Oftentimes the higher level LANs, reflecting the volumes of pooled traffic, use faster media, such as 16 Mbps token ring or fiber/copper distributed data interface (F/CDDI). ATM LAN switches, with their much higher speeds, are a natural in this role.

With the ATM Forum's LAN emulation, one gets a marriage of necessity between two very different partners. The LANs user shares the bandwidth with neighbors, communicates locally via media access control (MAC) addresses, and employs variable-length frames to package data. In the case of Ethernet, all messages are broadcast to all users on the LAN segment. The ATM user, by contrast, sets up an end-to-end SVC connection prior to transmission, employs ATM addressing, packages data in uniform 53-octet cells, and tears down the connection after the transmission is concluded. LAN emulation, often referred to as "LANE" (Phase 1), is the ATM Forum's quick solution to getting these two very dissimilar technologies to work together. (A more general solution, incorporating both ISO layers two and three, is in development and is called "MPOA," for multiple protocols over ATM.)

The LANE specification defines a standard way to correlate LAN MAC addresses with ATM virtual circuits. The LANE technology requires two specialized servers, the broadcast and unknown server (BUS) and the LAN emulation server (LES). The BUS and LES are typically implemented in software on a switch that connects the Ethernet ports to ATM.

The BUS and LES collect and store addressing information. When a workstation or server directedly attached to the ATM network—called a LANE "client"—needs to contact a node on the legacy LAN, it issues a request that is received by the LES. If the LES knows the address, it responds. If not, it signals the BUS to broadcast the query over all the ATM virtual circuits to all nodes on the network. When located, the node responds with its MAC address and the BUS forwards it and the associated virtual circuit to the LES. The LES, in turn, forwards the information to the LEC and also updates its tables, allowing it to respond directly and avoid unnecessary future broadcasts. And to improve efficiency further, the LECs cache the addressing information so they do not make unnecessary LES requests.

Several other aspects about the LANE specification need be noted. First, the switch performing the LANE function is also a LEC, so when an Ethernet node tries to reach another station across an ATM backbone, the switch figures out which virtual circuit to use. Second, LANE defines a way for LECs to set up virtual circuits to transmit data. Known as data direct virtual circuits, they employ UNI 3.0 or 3.1 standards. Third, LANE also describes control VCCs, which are used to transfer cell control information between the LEC and LES or BUS. Finally, LANE specifies the segmentation and assembly (SAR) functions that render Ethernet or token ring frames into ATM cells.

Overall, probably the most important single feature (or limitation) of LANE is that it represents a MAC layer bridging (ISO layer 2) solution, which means that a router is needed to connect emulated LANs. And this limitation portends the coming of MPOA.

Below is an illustration (Table 3.1), employing the ISO's seven-layer typology, but expanding the layer 2 data link shows where LANE functions in an Ethernet-TCP/IP environment.

Having described how LANE works and its place in the protocol stack, a not impertinent question remains: Does it work as described in the LANE 1 standard and can anyone demonstrate it today? Kevin Tolly in *Data Communications* magazine investigated in the September 1995

Table 3.1
LANE Functionality Within an Ethernet TCP/IP Environment

LAN	*ATM*	*LAN*
7 Application		Application
6 O/S-specific		O/S-specific
5 API socket		API socket
4 TCP		TCP
3 IP		IP
2 Ethernet MAC	Ethernet MAC	Ethernet MAC
	ATM LUNI	
	ATM AAL	
	ATM Layer	
1 Physical	Physical	Physical

Key: AAL: ATM adaptation layer; API: applications
programming interface; IP: Internet protocol; LUNI: LANE
user-to-network interface; MAC: media access control;
O/S: operating system; TCP: transmission control protocol.

Source: Ross, T. L., "ATM APIs: The Missing Links,"
Data Communications, Sept. 1995, p. 120. © 1995
The McGraw-Hill Companies. All rights reserved.

issue and found the answer was mostly no. Only FORE could demonstrate it and they relied on a number of proprietary features of their simple protocol for ATM network signaling (SPANS) software to do it [1]. In a return engagement by *Data Communications* a year later, 15 vendors were solicited. Six accepted: Bay, FORE, Cisco, IBM, Newbridge, and 3Com. Bay and Newbridge used proprietary LANE software, and Newbridge offered prestandard MPOA. Cisco's catalyst 5000-Lightstream 1010 combination and 3Com's Cellplex 7000 emerged as the top performers [2].

3.2 MAKING A CONNECTION

As noted earlier, making an ATM connection is analogous to making a telephone call. The switching resources have to be in place (signified by a dial tone). The circuit resources have to be in place (signified by the ring-ing). And the connection resources have to available and consenting (signified by the other party answering).

With ATM, a single SVC may traverse a half-dozen switches, which, conceivably, could be furnished by a half-dozen different switch vendors. These switches, link by link, must exchange information regarding connection QoS (CBR, ABR, etc.) and, once data transfer is in process, information on potential congestion via a "rate-based" congestion algorithm. As the congestion algorithm reflects a 1995 decision by the ATM Forum, most vendors have not implemented it, much less tested it in a heterogeneous environment. Today's commercial ATM networks provided by interexchange carriers (IECs) and competitive access providers (CAPs) (MFS Worldcom [ex-WiTel, ex-MFS Datanet], AT&T, British Telecom [ex-MCI], Sprint, Teleport) avoid these problems by using one, or at most two, switch vendors to provide what is in effect a subset of ATM services.

The actual back-and-forth communication process setting up an SVC is called connection admission control (CAC) and was depicted in the previous chapter in Figure 2.5. In this negotiation, the requesting node informs the network of the type of service required, the traffic parameters of the data flows and QoS for each direction.

The traffic parameters are as below:

- *Peak cell rate* (PCR): The maximum rate at which cells can be transmitted across a virtual circuit, specified in cells per second and defined by the interval between the transmission of the last bit of one cell and the first bit of the next.
- *Cell delay variation* (CDV) or *cell delay variation tolerance* (CDVT): Measures the allowable variance in delay between one cell and the next, expressed in fractions of a second. When emulating a circuit, CDV measurements allow the network to determine if cells are arriving too fast or too slow.
- *Sustainable cell rate* (SCR): Maximum throughput bursty traffic can achieve within a given virtual circuit without risking cell loss.
- *Burst tolerance* (BT). In the context of ATM connections supporting VBR services, it is the limit parameter of the generic cell rate algorithm (GCRA) whereby conformance to the traffic contract is defined. The GCRA, in turn, is defined by two parameters, the increment (I) and the limit (L).
- *Minimum cell rate* (MCR): An ABR service traffic descriptor, in cells per second, that is the rate at which the source is always allowed to send.

The result is that CAC, seemingly straightforward as depicted in Figure 2.5, becomes a complex, three-dimensional (Service × Traffic

Table 3.2
Service Classes and Applicable Parameters

Attribute	ATM Layer Service Categories					Parameter
	Constant Bit Rate (CBR)	Variable Bit Rate (VBR) Real-Time	VBR Nonreal-Time	Available Bit Rate	Unspecified Bit Rate	
Cell loss ratio (CLR)	Specified1	Specified1	Specified1	Specified2	Unspecified	Quality of service (QoS)
Maximum cell trans-fer delay (CTD) and mean cell transfer de-lay (CDV)	CDV and max CTD	CDV and max CTD	Mean CTD only	Unspecified 6	Unspecified	QoS
Peak cell rate and cell delay variation tolerance	Specified	Specified	Specified	Specified4	Specified3	Traffic
Sustain-able cell rate and burst tolerance5	N/A	Specified	Specified	N/A	N/A	Traffic
Minimum cell rate	N/A	N/A	N/A	Specified	N/A	Traffic
Congestion control	No	No	No	Yes	No	

Source: Alles, A., *ATM Internetworking*, San Jose, CA: Cisco Systems, 1995, p. 55.

Notes:

1) For CBR and VBR the CLR may be unspecified for CLP = 1.

2) Minimized for sources that adjust cell flow in response to control information.

3) Not subject to call admission control and usage parameter control procedures and may use different value from Section 3.6.2.4 of the User Network Interface 3.1 specification.

4) Represents the maximum rate at which the source can send as controlled by the control information.

5) These parameters are either explicitly specified for permanent virtual circuits or switched virtual circuits as defined in Section 3.6.2.4.1 of the User Network Interface 3.1/3.0 specifications.

6) The objective of the service is that the network does not excessively delay the admitted cells. Requirement for explicit specification of the CTD and CDV is for further study.

Parameters × QoS Parameters) metric. Table 3.2 below reduces this complexity to two dimensions; note that a number of combinations are still not specified.

3.3 MULTIPLEXING AND DEMULTIPLEXING

Everyone is familiar in nature with how several small streams of water combine to form a swift larger one, and, conversely, how a large stream will divide into many small tributaries. The practice of combination for efficient transit and decombination for efficient distribution has been a fixture of modern telecommunications for decades. Most are familiar with the voice network "mux" and "demux" hierarchy of the public switched voice network. The basic customer unit ("stream") is the 64-Kbps (DS-0) time division multiplexed voice circuit. Narrowband data circuits were created by digital access controllers (DACs) that demuxed the 64-Kbps streams further, usually in various increments down to 4.8 Kbps. In the other direction, large numbers of 64-Kbps circuits were "muxed up" or "rolled" into larger rivers by "channel banks," 24 64-Kbps circuits into standardized DS-1s and further, 672 into proprietary DS-3 formats.

The down sides of this muxing up and muxing down were that it involved considerable hardware and the hardware tended to make the process inflexible and greatly slowed circuit provisioning, often taking a calendar month or more to install a data line of a certain capacity. With increased use of software control, the provisioning cycle has been lessened considerably, particularly in urban areas where the vendors have most of their resources (and customers) concentrated. The broadband dream of bandwidth on demand is through sophisticated software control, smart ATM switches, and enormous residual capacity in the vendor's fiber-optic-dominated transmission network, the customer—not the vendor—will be able to instantly provision what is required when a particular application wants it.

Economically, this vision also subsumes a fiber-optic version of "all roads leading to Rome." Indeed, a number of smallish communities have made considerable investments in communications infrastructures (via direct carrier subsidies) in order to attract (or retain) large teleservice or telemarketing operations. Many communities view their location on the "information highway" similarly to having harbor, bus, rail, and jet service. Nevertheless, in the absence of local, state, or federal subsidies, such vendor facilities must be cost-based and the resources available (links and nodes) in any particular location are ultimately finite. As more and more of the telecommunications market is deregulated and cost-based, huge service area inequities are likely.

Uncertain at this writing (Summer 1996) are what will be the popular sizes of bandwidth corresponding to the old electrical hierarchy's DS-0/1/3. The optical hierarchy provides an immense spread:

- OC-1 (51.84 Mbps);
- OC-3 (155.52 Mbps);
- OC-12 (622.08 Mbps);
- OC-24 (1.2 Gbps);
- OC-48 (2.4 Gbps).

It is safe to assume that the small sizes will be the most popular, but the decisions made by the cable and satellite providers of video to the home will greatly influence the equipment market and accelerate the popularity of particular channel sizes, just as the standardization of DS-1 (also known as T-1) did for that market earlier.

Whatever the channel size, ATM can be viewed as the latest and most sophisticated of statistical multiplexers. "Statistical" multiplexing is when several streams are combined on a channel in such a manner that the aggregate of the streams (such as bursty data traffic) can temporarily exceed the channel-carrying capacity without data loss. The burstiness is handled by temporarily buffering (storing) the excess. This, of course, adds to the transmission delay. In severe cases of excess offered traffic, conventional statistical multiplexers quiesce one or more sources. ATM allows what one might call a hybrid solution. Like conventional multiplexers, it allows for a protected partition for high-priority traffic (CBR, VBR) while statistically multiplexing lower priority traffic (ABR, UBR). This is illustrated in Figure 3.1. Because of the high bandwidth/data volumes involved, ATM also incorporates the option of selectively discarding low priority traffic that threatens to overwhelm the switch, as well as quiescing or throttling down the sources.

Last, mention must be made of ATM service multiplexers. Service multiplexers combine and decombine different services (voice, computer data, video) into one or more high-bandwidth ATM streams (pictured in Figure 3.1). They can be viewed as a special case of an ATM edge switch; their conventional telco counterpart is the digital access controller (DAC) or channel bank. Although the service multiplexer role seems straightforward in theory, it is almost never so in practice, as digital service units (DSUs) and specialized processors for encryption/decryption, compression/decompression, silence suppression/restoration, format or protocol conversion, and so forth are almost always present as well.

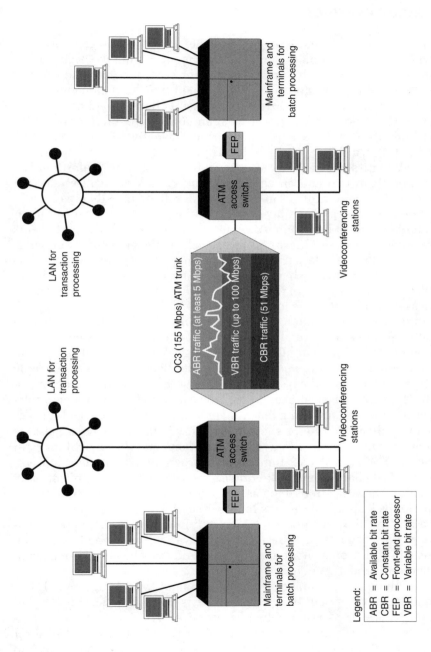

Legend:

ABR = Available bit rate
CBR = Constant bit rate
FEP = Front-end processor
VBR = Variable bit rate

Figure 3.1 Statistical multiplexing. (*Source:* Hughes, D., and K. Hooshmand, "ABR Stretches ATM Network Resources," *Data Communications*, April 1995,   p. 125. © 1995 The McGraw-Hill Companies. All rights reserved.)

3.4 ERROR CONTROL

In music, when one or more players flub their parts—make gross errors—the drill is to "take it from the top," replay the piece from the beginning of a page of music. Similarly, present day error control usually consists of one or more retransmissions over the affected link. In data communications, the retransmission unit (frame, packet, cell, message) can vary as current layered protocols perform what amounts to redundant (or multi-layer) integrity checking. Nonetheless, most usually it is the data link layer (ISO 2) with its cyclical redundancy check (CRC) that occasions these retransmissions. Usually it is invisible to the user. Unfortunately, this is not always so and an egregiously noisy link can cause constant retransmissions and throughput degradation, called "throttledown."

Note, too, that present day error control is usually based on a link or portion of an end-to-end connection. One is familiar with the party game where some nonsense phrase is successively whispered from one guest to another. After very few transmission over an error-prone link (guests whispering), the phrase, whatever its initial understandability, is reduced to gibberish. The fragility of the individual link relative to the integrity of the message is the rationale for link-by-link error checking.

In data communications one can well imagine the effects of pathological conditions on several portions of an end-to-end connection. The throttledowns become cumulative—and large. Various weird things can happen. When the end-to-ends are highly intelligent, such as two people on a transoceanic call, delays can be highly variable and the communication (technically, "user to user") can proceed successfully. When a telephoner reaches an answering machine or voice mail service (technically, "user to application") he or she can decide whether to leave a message or hang up. Other applications, such as automated teller machines (still another "ATM"), have a set limit to their programmed patience and will hang up. Most affected by throttledowns are computerized "application to application" conversations such as when a retail cash register autodials a credit or check approval service. In this case, the throttledown may cause a time-out, yet the cause of the (apparently) negative result may be ambiguous. For instance, a little old lady, invariably somebody's mother buying a birthday present for a new grandchild, may be perceived a deadbeat and erroneously pinched.

As alluded to above, error control is one of today's traditional data communications services that many believe is anachronistic. Critics point out that where DS-1 based networks promised error rates of $10-7$, networks based on synchronous optical networks (SONET) advertise error rates in the $10-11$ range. These critics argue (1) that error control has no more place on today's fiber networks than horses on superhighways, and furthermore (2) error control should be solely end-to-end, at the "edge of the network" rather than link-based.

More cautious commentators concede the truth of the lower error rate, but point out that with the increased data volumes (in the plumbing analogy, the increased diameter of the "pipes") the errors will occur just as frequently. They also note that dropped or switched bits may have out-sized effects on some applications, such as financial transfers or where the data is encrypted. In the short run—which may prove to be decades—industrial-strength error control (16- or 32-bit CRC) will remain a feature in high-speed data networks. Why? Because most of the data traffic passed over the broadband networks of the future will be encapsulated present day network protocols, and virtually all use 16/32 CRCs. As Ronald Reagan once rationalized his B-movie film career, "They didn't want art, they wanted it Thursday."

Having said that ATM will typically be transporting, via encapsulation, the error-control schemes of predecessor higher level protocols, how does ATM police its own act? ATM's error control is found in two general locations: the ATM header and as a component of ATM AALs. One octet of ATM's 5-octet header is dedicated to a header checksum. Error checking the AALs is summarized below:

- AAL 1: AAL 1 adds a 1-octet header to each data block; this header contains sequencing, timing, and error-correcting information. Single bit errors can be corrected if required.
- AAL 2: Neither the ITU-T nor ATM Forum have defined the specific format of AAL's segmentation and reassembly (SAR) protocol data unit (PDU). The best prognostication is that AAL 2, if eventually specified, will be similar to AAL 1 and will include sequence checks and will detect/correct single bit errors.
- AAL 3/4: AAL 3/4 adds a 2-octet SAR header (which contains a sequence number) to each block. This allows it to identify misinserted or lost packets. It also adds a 2-octet trailer that contains a length field (which allows it to count data field octets) and an error-correcting code to detect and correct single bit errors or detect multiple bit errors.
- AAL 5: AAL 5's SAR layer uses a single bit in the ATM header at the end of each SAR data unit. The AAL 5 convergence sublayer (CS) adds an 8-octet trailer in the last cell. This trailer has four fields that include a 2-octet length field and a 4-octet cyclical redundancy check (CRC). The CRC is capable of detecting multiple bit errors and misordered cells.

3.5 LATENCY

Latency is media delay. In cooking, there is latency between when one turns on the gas flame and when the fat sizzles in the fry pan. In high-speed network switching, with the potential at least for some links of relatively long segments of high-capacity fiber-optic media, dealing with latency entails provisioning large buffers. Although the large buffers induce a delay in themselves, they allow the dampening of its effects (technically, the delay variation). The buffer space required equals throughput x times round-trip delay y, the "delay bandwidth product." In round numbers, this amounts to about 1 KB of buffer per kilometer, and about 1 GB of buffer per second [3]. Dealing with the maximum product necessitates a plan for buffer allocation (a series of independent buffers versus shared memory) and management strategies for handling congestion.

3.6 DATA PACKAGING

Packaging for data transmission (files, messages, frames, packets, cells, etc.) is directly analogous to the various containers that fill our spaces and litter our dumps. Similar to their corrugated and styrofoam-filled cousins, they vary by expense (ratio of data bits/packaging), layers of wrapping (protocol layers employed), security (encryption), tamper resistance (CRC), ease of handling (addressing), and reliability of delivery (retransmission).

One observation about packaging is that it has to be periodically rethought or reinvented. To take an example from the auto industry, in the beginning imported cars were treated as general cargo, stored in the holds, and winched out of the holds to dockside. Nowadays, the most efficient importers use custom-designed, drive on, drive off ships that look like floating, elongated parking ramps.

In today's high-speed data transmission employing ATM, the data is typically encapsulated, preserving the current protocols, usually to preserve the addressing scheme of an extant application. (Managing this encapsulation process in ATM are the "adaptation layers" or AAL.) Thus one encounters use of the U.S. Department of Defense's transport control protocol/Internet protocol (TCP/IP), the protocol equivalent of a humvee, being transported (to continue the metaphor) by the network equivalent of a French Train à Grande Vitesse (TGV). From an efficiency point of view, the ratio of overhead octets to data octets, it is not a pretty sight. On most applications where encapsulation is employed, the efficiency is certainly lower than that of 1960's era asynchronous ASCII, where one wrapped a

7-bit code with a start bit, stop bit, and parity bit, achieving somewhat less than 70% efficiency, as some of the data were control instructions.

Prospects for new applications are much brighter. Faced with the requirement of delivering a steady stream of compressed video as economically as possible, and with error detection/correction a nonissue, ATM-delivered video will likely employ an AAL 2 standard optimized for the service.

Relative to a judgment on the efficacy of ATM's data packaging, several observations are germane. First, except in the case of the massive forward error-checking protocols used in some wireless applications, protocol overhead is the price one pays for applications flexibility and ATM is arguably the most flexible of current packaging schemes. Second, ATM package overhead was intended to be service-specific, with AAL 3/4 the highest and AAL 5 the lowest, but ultimately adjustable. For example, nothing precludes voice traffic from being carried VBR-RT or even ABR. Third, the ubiquity of broadband transmission makes bit and octet counting arguably less significant in the equation than formerly. Fourth, the lower level (ISO 2/3) protocol inefficiencies are often more than offset by presentation (ISO 6) efficiencies, particularly with compressed video where compressions of four and three to one are achievable without discernible impact on the application.

3.7 PERFORMANCE

High performance stems from fast components working together. The drop-nosed orange French TGVs not only have powerful turbine engines but computerized speed and suspension systems, and are supported by advanced switching systems, precise track alignment, and high-quality roadbeds. In today's electronics milieu, silicon and memories and buses offer, at a given time, certain price/performance ranges. Very high performance is available only at very high cost. As consequence, one tries to create a product where high performance is achieved by a combination of marketplace components and clever design.

The most common of these design tricks is parallel processing. Most are familiar with the concept of a bucket brigade, where buckets are handed back and forth along a single line. Communications people would characterize the typical bucket brigade as a "full-duplex, serial channel." The process could be speeded up incrementally by faster workers (components) or by having the buckets passed in a single direction (half duplex). The throughput, however, would scale linearly if another 10 bucket brigades were added in parallel. This is the general idea with parallel designs, near-arithmetic scaling with near-arithmetic cost increases. And this design is particularly apt for high-speed communications where there may be multiple logical flows (four channels video, six channels computer data) sharing one physical channel.

The fastidious, however, will have noted the caveat "near" regarding both performance increases and costs in regard to parallel designs. Parallelism, even the most clear, always requires overhead processing. In the case of cell throughput, most high-speed switches require initial sorting in order to assign logical flows to processing streams. Where a logical flow may be split between two processing streams, recombination and ordering may be required. In the case of cost, common switch components such as power supplies and card cages and buses pose capacity limits beyond which costs rise nonlinearly; real-time control software brings its own limitations. In sum, parallel design employing merchant electronics that linearly scales by cost/performance, and that can be neatly packaged in popular increments (usually the powers of 2), is everyone's dream—and no one's reality.

With ATM, high performance is probably the single driving feature, even more than its facility to handle multimedia, deliver QoS, or provide a single protocol architecture for LAN/MAN/WAN environments. This emphasis, particularly on the high-speed transfer of computer data, can be expected to continue. As noted later in the vendor tables (Tables 5.1 and 5.2), few vendors provide OC-24 (2+ Gbps) switch interfaces today and few Bell Operating Companies (BOCs) and interexchange carriers (IEXC) offer transmission services at that speed. But, in a telling example of the role performance plays, the recently issued (8/23/95) draft Request for Proposals (RFP) for Post-Federal Telecommunications Services 2000 (PF2K), which reflects a compendium of United States Government requirements, requires responsive vendors to be able to provide ATM services up to and including OC-48 over the continental United States and to specified foreign locations beginning in 1999.

3.8 SEQUENCING

Many have found themselves at a children's Halloween or Christmas celebration that concluded with the tots lining up on stage, each holding a letter card, intended to spell out a greeting to the loyal parents, and hilariously mis-sequenced. In data communications, sequencing is one of those tasks that everyone agrees is necessary but not everyone agrees on who should do it. It would be most disconcerting were video or voice transmission presented to the viewer or listener out of sequence. Indeed, such "flashbacks" or "voice-overs" are common tricks of the film maker. Nor is sequencing necessarily real-time. One is familiar with common commercial transactions such as credit or balance checks (usually interactive) prior to withdrawals (usually batched).

Most current commercial transport networks accept sequencing as part of their task. Sequencing requires serial numbers, at least occasional sorting, and often waiting for retransmissions. Network minimalists hold that sequencing, as well as other grubby network administrative work, is inappropriate for a network whose focus should be the speedy transmission of billions of bits. They view sequencing the way some view bicyclists on freeways. They maintain that these tasks would be more appropriately accomplished beyond the network endpoints. Further, they point out that most network protocol stacks are full of what they see as petty, efficiency-robbing redundancies. Often they are correct. Network sequencing, an ISO level-three function, is almost always repeated at level seven in application areas like financial services or messaging systems. In God's eye, such duplication is inefficient; pragmatically, however, today's network consumers expect a sequenced product and are willing to put up with, and pay for, a good deal of inefficiency in communications. Moreover, were it otherwise—should their data arrive unsequenced—some would have to redesign their applications.

ATM virtual circuits, unlike the Internet, operate logically in a first-in, first-out (FIFO) manner. Although physically there is both service multiplexing (voice, video, computer data) and VC multiplexing (computer data 1, computer data 2, ...), the data are still FIFO. In the case of discarded cells under ABR, where retransmissions are necessary to restore a broken sequence, resequencing is performed, but at the packet level by the transport protocol (TCP, TP4) and not at the cell level.

3.9 BUFFERS AND BUFFERING

Buffers are life's waiting rooms. Just as the patients (presumably, the inexpensive resource) sit about reading old magazines in a physician's office so that access to the physician (the expensive resource) can be optimized, so it goes in automata. Given their usual connection with delays, buffers are viewed as a necessary indulgence to a world filled with traffic surges, speed mismatches, requirements for error correction, and other pesky but real-life concerns.

High-speed networks, however, have opened a whole new set of questions regarding buffers. Advances in memory technology have relatively recently allowed (by historical standards) huge buffers for small cost. Network processors of today can economically offer buffers that exceed the main memory of mainframes in the recent past. And symmetrically, with high-speed networks, much is expected of them. One author has presented it as follows:

> "A single-source buffer calculation for an OC-3 (155 Mbps) ATM connection spanning 3,000 kilometers (about 1,850 miles would be as follows:
>
> Bandwidth = 353,208 (OC-3 delivers 353,208 cells per second) Round-trip transmission time = 60 milliseconds (this is calculated by multiplying the round-trip connection distance—in this case, 6,000 km—by the speed at which light travels through a fiber optic cable, which is 5 km/ms.) 353,208 × .06 = 21,192 cells" [4]

And, at least in theory, all or a portion of these 21,912 cells might be received in error and have to be retransmitted. The conventional wisdom is to bemoan the size and cost of these buffers. Others have pointed out that relative to the cost of the transmission pipe—itself a kind of buffer—huge modern buffers are the cheap solution to many of the pesky problems noted.

3.10 QUEUES

Queues are the civilized way of waiting. One encounters them everywhere. Most banks impose what is technically a "single queue" (line) and "multiserver" (several cashiers) arrangement. Small airlines, particularly at small airports, often offer a single queue and a single server. Large airlines at international airports generally present a multiqueue-multiserver arrangement.

Everyone at one time or another has encountered blockage, or a queue with (seemingly) infinite holding time. In the communist Poland of yesteryear, for instance, a long line at shop was called a "sklep w ogon," or "shop with a tail," humorously characterizing consumer behavior in the face of scarcity. Those familiar with third-world telecommunications environments have their pet stories, funny after the fact, of waiting for dial tone, and once dial tone was obtained, repeatedly dialing a rotary phone seeking to obtain an intercity trunk. Reaching the desired party often became anticlimactic.

In high-performance environments, a particular queuing problem called "head of the line" blockage is important to avoid. Technically, it is akin to "page faults" in virtual memory operating systems and the "fatal embrace" in database systems; it involves a processing stoppage as a result of resource contention. For example, one is waiting in a bank queue and the teller cannot complete one's transaction without the authorization of a bank officer, who is herself busy.

Although in the past these sorts of contention problems have been cleared by timers—analogous to a dedicated bank officer to clear difficult cases—the modern approach has been to proliferate paths. Aided by low-cost digital electronics, the incremental cost of parallel queues is small when compared to additional WAN diversity, or worse, the unavailability of a highly visible application or service.

Everyone hates queues, and, in a sense, broadband communications represent one more attempt at achieving the perception of "queuelessness"—that somehow we will dispose of a huge number of bits in a magic pipe and we shall get everything we transmit instantly. As with other schemes, ATM will disappoint. Almost any cost-based scheme will involved class-of-service strictures (discussed below) wherein low-use time, often late night or early morning, will be bargain priced. In this scheme, any discount traffic that arrives prematurely, like the too-early party guest, will be queued and/or back-pressured (quiesced). Given the potential volumes, the caveats above on buffer consumption also are relevant: a switch, to protect the integrity of its buffers, must be exceedingly nimble at either refusal (quiescing the source), buffer and quiesce, or (worse case) buffer/overflow/quiesce/retransmit.

As noted, ATM's technical approach is to minimize unnecessary queuing delay. However, a permissive entry policy moves the problem downstream into the ATM nodes and links. As ATM congestion control consists of a number of elements and is quite complex, it will be overviewed in the subsequent section.

3.11 CONGESTION CONTROL AND FLOW CONTROL

Everyone who has tried to make the 8 p.m. Friday night show of a new and popular movie knows about congestion. Too many customers overwhelm the ticket booth (the admission control function) and, perhaps, there are too many to fit into the theater (the transmission buffer function). Flow control in the movie industry usually means portable stands with wind-out nylon belts to enforce a single or multiple server queues for admission, often in conjunction with back-pressure and informal counting to limit physical occupancy to the maximum permitted by the fire marshal. Occasionally, large numbers of customers are turned away at the ticket booth. Depending on the ticket policy, sometimes people with tickets are not seated, are recycled to subsequent shows, or are refunded their money (although a "half-duplex" illustration with QoS, congestion, and flow control in telecommunications is pretty much the same).

Congestion control is probably the most humbling of telecommunications subjects. Communications engineers are brought up on the sequence: study the behavior, understand the behavior, automate the behavior. That important aspects of the First-World infrastructure work reliably (gas, electricity, telephone) serve as a testament to the efficacy of this credo and its statistical underpinnings. Nonetheless, one has come to periodically expect "spikes" in mass behavior that regularly lay low portions of the telecommunications fabric. For instance, in 1994, the Ottawa radio station CHEZ-FM offered 53 pairs of Pink Floyd concert tickets free to callers. An estimated 300,000+ call attempts overwhelmed a specially equipped exchange, causing delayed dial tone and other problems in an area of over 100 miles [5]. The problem will not go away, nor is it limited to public networks; if anything, rapid increases in televised direct sales, electronic "town meetings," and home shopping channels promise new excitement.

In theory, all events, even in surge conditions, should generate routine responses. If the long distance circuits do not have adequate capacity on the primary path, the excess demand should be routed onto secondary paths and the calls should go through. If the resources available for rerouting are temporarily exhausted, one should get a recording saying so. So, too, with local exchanges experiencing surges; they should provide a recording announcing that all circuits are temporarily busy and directing one to call later. This is what is supposed to happen.

Unfortunately, this is not what happens all the time. The 1981 AT&T divestiture made competitors out of cooperators, postdivestiture "startup" companies are now billion dollar corporations, and telecommunications use has greatly increased. New services (mobile phone, mobile data, pagers, call waiting, messaging) have appeared and others (fax, dial-up data) greatly increased. No one—not AT&T, the Pentagon, the Federal Communications Commission, the 50 state regulators—even pretends to effective control. Ultimately, only the market, where you cannot successfully bill for a service that does not work, provides a discipline to this happy chaos.

There several reasons why contemporary efforts at congestion control have become less tractable. Some believe that our centralized models for handling congestion control are simply inappropriate to a semianarchic, distributed universe. For example, a marketing company using brand x equipment contracts with a network supplier using y equipment to carry z amount of data traffic. The company experiences a stunning response to a new service and generates z (cubed) traffic. The software configuration profile controlling congestion between brand x equipment and brand *y* equipment has either been improperly set or is otherwise inconsistent and *z* (cubed) overwhelms a portion of the carrier's network, in the process denying service to other customers.

This would seem a trivial problem were it not that there are dozens of switch types with millions of lines of control software. In the course of a year, it is likely that this software will be stress-tested at thousands of locations by millions of customers with billions of accesses. Bugs or inconsistencies will appear. Human beings estimate traffic and configure the switches. Despite highly detailed standards and rigorous control and testing, the process is prone to specification errors. Weird things cannot only happen but the resulting pathologies are often propagated (e.g., one switch passes incorrect information to neighboring switches). In sum, historically, perfect hardware/software configuration control and switch management was unattainable in a highly homogeneous and hierarchical environment with relatively few traffic types (predivestiture AT&T). It is now much more difficult with a heterogeneous, distributed environment with many more players, with many more traffic types, and much higher volumes.

Congestion and flow control are among the most complex issues with the ATM technology. The reasons for this complexity are readily evident. First, ATM purports to function as a *service* multiplexer, harmoniously combining voice, video, and computer data according to QoS guarantees. Second, ATM's services, ABR and UBR in particular, are statistically multiplexed, which implies input flows, however brief, exceeding channel capacity. Third, ATM is a high-speed cell-switching technology where delays of any kind (queues, buffers) are to be avoided at all costs. Fourth, although ATM switches can protect themselves by selective discarding of ATM cells, the requirements of many data types, as well as the additional complications introduced by encryption and/or compression, make this option very unattractive from a throughput standpoint. Fifth, ATM is a so-called "unification" technology and will appear in LAN, MAN, and WAN environments where the fielded equipment will have large differences in capacity, particularly in buffering.

This complexity was much reflected in disagreements within the ATM Forum. In the Forum's recent decision in favor of "rate-based" flow control (where all switches in the connection path agree on incremental throughput adjustments to ABR/UBR flows), the merits of a competing, credit-based proposal were strongly argued. (With "credit-based" flow control, all switches in the connection path agree on the traffic adjustments on the basis of available buffer capacity.)

It should be pointed out early in any discussion of ATM congestion and flow control that today's ATM switches do not employ standard approaches. Today's public ATM networks are based on homogeneous switches using proprietary (often termed "prestandard") techniques. ATM networks employing switches from several manufacturers often are forced to rely on PVC connections and network interface cards (NICs) that limit input flows; others simply overengineer their networks and sacrifice statistical multiplexing.

With the various warnings and disclaimers past, what are the range of traffic and congestion control functions available to ATM? Basically there are nine such mechanisms, though (as explained below) no ATM switch manufacturer makes use of all them; they are as follows:

1. Connection admission control (CAC);
2. Usage parameter control (UPC);
3. Selective cell discarding;
4. Traffic shaping;
5. Explicit forward congestion indication (EFCI);
6. Resource management using virtual paths;
7. Frame discard;
8. Generic flow control (GFC);
9. ABR flow control.

CACs are the actions taken at SVC establishment to determine whether a connection can be "progressed" or not. To use a transportation metaphor, it is akin to a truck driver having his passport, license, registration, insurance, cargo permits, transit visas, and so forth in order prior to departing on an international run.

UPC, also known as traffic policing, is directly analogous to state troopers enforcing an automobile speed limit. In the case of ATM, it means that the ATM switch must enforce a particular connection-based traffic contract. In practice, this means limiting the peak rate of a connection to that of the slowest link along the path; the automobile analogy might be the reduced speeds encountered with heavy traffic.

Selective cell discarding is just what it implies—the ATM switch throwing away premarked cells considered disposable. An approximate transportation example would be, in emergencies, the banning of automobiles from the central city except for police cars, taxi cabs, and those vehicles marked "For Official Use Only." In the case of ATM, it consists of discarding cells whose cell loss priority (CLP) equals 1, so to protect, as long as possible, CLP = 0 flows.

Traffic shaping (or pacing) refers to a mechanism that alters the characteristics of a stream of cells on a connection so to better meet QoS objectives. An automobile analogy to traffic shaping is the stoplights that one finds on some freeway entry ramps; they regulate (delay) access with the objective of maintaining a minimally acceptable speed on the freeway itself. Traffic shaping in ATM is limited by being required to preserve the cell sequence integrity of the connection. In practice, shaping almost always increases delay, measured here as the mean cell transit delay (CTD).

How and where in the network shaping is implemented is network-specific. The ATM Forum's UNI 3.1 specifies a dual "leaky bucket" mechanism for UPC and shaping can be visualized as the "front-end" of policing (Figure 3.2).

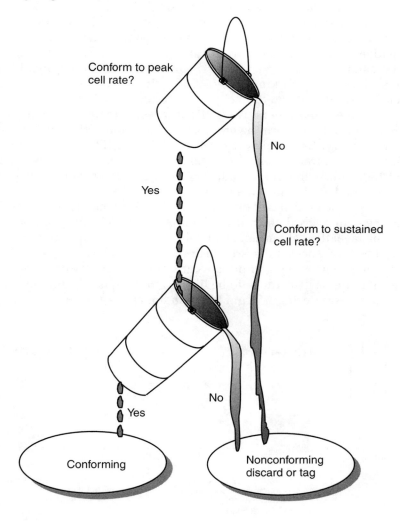

Figure 3.2 "Leaky bucket" mechanism to implement usage parameter control. (*Source:* Modarres, H., *Performance Considerations, Wide Area ATM Networks*, Herndon, VA: Newbridge Networks Corporation, 1995, p. 7.)

A "leaky bucket" mechanism implies a finite-sized bucket into which traffic flows and a maximum flow rate or "leak" from which traffic emerges. In ATM terms, the size of the bucket determines the cell delay variation tolerance (CDVT); the leak rate is the peak cell rate (PCR). Details of these algorithms are discussed at great length elsewhere [6,7]; suffice it to say that the effects of this electronic leaky bucket are to "shape" or "pace" traffic in the manner depicted on the left side of Figure 3.3.

Explicit forward congestion indication (EFCI) is a cell header setting that may be examined by the destination end system. Its highway analogy is the freeway signs that warn motorists of impending delays and possible alternative routes. The highway parallel is particularly apt because the ATM switch, like the tired motorist, may choose to do nothing. EFCI is optional for CBR, both kinds of VBR and UBR; only in ABR is the end system obligated to do anything upon receiving the EFCI.

Resource management using VPs allows one to implement a form of priority control by segregating groups of virtual connections according to service category. Essentially, it consists of intelligent planning or, more baldly, plain overengineering to avoid congestion. A highway parallel might be a toll road running adjacent to a free road where congestion could be avoided, literally, at a price.

Frame discard is based on the reality that if a congested network element needs to discard cells, it is oftentimes more effective to discard at the frame level than at the cell level. Often this shows up in vendor literature as the "early packet discard" feature. Where cell discard was likened to the emergency banning of private car traffic from a city center to avoid congestion, frame discard could be likened to the special banning of large trucks and buses during AM and PM rush hours.

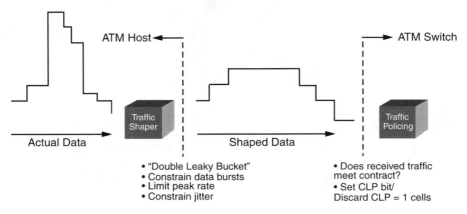

Figure 3.3 Traffic shaping and policing. (*Source:* Alles, A., *ATM Internetworking*, San Jose, CA: Cisco Systems, May, 1995, p. 56.)

Generic flow control (GFC) applies only to ABR service and is "not precluded" at the UNI for draft Traffic Management Specification 4.0. It is not expected to be implemented, but remains a tool in ATM's conceptual toolbox for congestion control.

In ABR flow control, the source adapts its rate to changing network conditions. It can be likened to the ultimate "smart" highway where sensors and controls operate dynamically, raising and lowering the speed limits, advertising stoppages, and maximizing throughput. One can appreciate the complexity of this process. In the ATM Forum the debate between the credit- and rate-based constituencies was very vigorous, with the rate-based approach the apparent winner. The potential payoffs of ABR flow control are appropriately large: where all CBR and most VBR traffic is expected to use fixed bandwidth portions, ABR is likely to be home for most computer data traffic—and the source of statistical gain in the network. The mechanisms that make ATM's ABR flow control possible—conveys state information about network elements to the source—are special control cells called resource management cells RM-cells).

3.12 SUMMARY

As noted earlier, the technical challenges posed by high-speed networks are numerous—too numerous for a single chapter. In ATM, these attributes have been further disaggregated into a huge number of terms, each with its forbidding acronym (see Acronym List). At the extreme, they allow the ATM cognoscenti to converse in ATMTalk, a language combining the honorific and soporific, and often opening the door to pecuniary gain. But the concepts behind the attributes discussed in this chapter—interoperability, connections, muxing, error control, latency, packaging, performance, sequencing, buffers, queues, and congestion and flow control—are familiar. Most get a dose of them each morning driving to work.

End Notes

[1] Tolly, K., "In Search of ATM LAN Emulation," *Data Communications*, Sept. 1995, pp. 29–30.
[2] Mandeville, R., and J.T. Johnson, "Forget the Forklift," *Data Communications,* Sept. 1996, pp. 120–134.
[3] Katevenis, M., Stefanos Sidiropoulos, and Costas Courcoubetis, "Weighed Round-Robin Cell Multiplexing in a General-Purpose ATM Switch Chip," *IEEE Journal on Selected Areas in Communications*, Vol. 9, No. 8, Oct. 1991, pp. 1265–1279.

[4] Hughes, D., and Hooshmand, "ABR Stretches ATM Network Resources," Data Communications, April 1995, p. 126.

[5] Neumann, P. G., Moderator, "Risks to the Public in Computers and Related Systems," in ACM SIGSOFT, *Software Engineering Notes*, Vol. 15, Issue 70, (also available via http://catless.ncl.ac.uk/Risks/15.70.html).

[6] Alles, A., *ATM Internetworking*, San Jose, CA: Cisco Systems, May 1995.

[7] Partridge, C., *Gigabit Networking*, New York, NY: Addison-Wesley, 1994.

The Technical Challenges of High-Speed Switching II

4

High-speed switching and, more largely, high-speed communications systems, depend on a large number of elements working in reasonable concert. Given general-purpose components, nothing is ever quite optimal, efficiencies and inefficiencies offset each other, and the system still works adequately for a particular application. Ultimately, the application's special needs—for security, for assured delivery, for seamless video—determine the relative importance of the network elements combined. Inescapably, there is a relativistic aspect when discussing them, or, as in this volume, even in the order in which they are presented. With this is mind, this second chapter on the technical challenges of high-speed switching again examines the various elements involved and how the ATM technology deals with them.

4.1 COMPRESSION

Anyone who has ever stomped down the trash in order to get the lid on has grasped the basic idea of compression. Compression is particularly important to the rapidly growing graphics and video areas and probably essential to any scheme of delivering the World Cup—or more plebeian fare—to the world's multitudes in a cost-efficient manner.

Digital compression techniques come in two flavors, lossless and lossy. Lossless compression is like sitting on a overfull suitcase to get it closed—hopefully, nothing is lost in the process. Lossy compression is like holding a garage sale—equally hopefully, the superfluous is reduced while the essential remains.

Currently analog TV transmission requires about 45 MHz of bandwidth for one-way broadcasts. Today's digital conversion and compression schemes reduce analog TV to DS-1 (1.544 Mbps) levels with decent fidelity and several lower increments with much less quality. For higher quality, the Motion Picture Experts Group (MPEG) II standard results in quite good compressed video at about 4 Mbps.

Looking forward, it is clear that a high-definition standard is forth-coming and will replace the U.S. NTIS standard (which dates from 1936 for black and white and was modified in 1952 to accommodate color). The new displays will be the technical cousins of the expensive, mega-pixel color displays often found with today's high-end workstations. The technical challenge posed by the new high-definition TV standards group is formidable: to achieve high-quality picture and sound; to avoid the ne-cessity of commercial channel reallocation; to coexist with NTIS; and, if possible, have a standard that unites Japan, the United States, and Europe, the geographical origins of today's three incompatible formats.

Most observers believe that several new formats will be blessed and will result in digital channels in the 6–8 Mbps range. Popular video soft-ware (Arnold Schwarzenegger epics, etc.), computer images, and video-conference outputs will be available, as the phrase goes, in a number of "popular sizes" along a quality-price continuum. Although a number of standards, from MPEG and the NTU-T, exist today, they have been under-cut by constant improvements in video compression technology. The Bal-kanized market among the content providers (Time-Warner, Disney), hardware providers (Sony, Toshiba, Intel), and distribution providers (Microsoft, DirectTV, Tele-Communications) adds additional complexity and necessitates alliances, often fragile, among giant corporations.

4.2 COST

Once upon a time, before software was born, one could intuitively com-prehend the costs of things. Matriculating from a Hershey bar to a used Ford convertible, one could grasp purchase costs, maintenance, licensing and fees, and insurance. Among grim accountants, there were further, more abstract measures—cost of money, depreciation, net present value, and worse. But then came software. Steve Jobs, creator of the Apple Macintosh, had a vision that software would enable simplicity. He was much enamored by the Cuisinart food processor, and, indeed, at some distance the early Mac and Cuisinart resemble each other. But the world has largely gone the other way, from "IBM-compatible," to "Wintel" PCs and more and more complex software and configuration options. Industry analysts say that the average yearly support costs of a business-use Wintel PC are approximately three times its (combined hardware & software) purchase cost; five times if it is connected to a network.

In the case of the most popular low-end ATM switches, they look like monitorless PCs. (The early FORE models even included a Sun Sparcstation in the box!) The choices are numerous below $25K and prices are falling. But the level of software complexity, the need to constantly upgrade that software, and the bewildering configuration options will make the costs of ATM ownership extremely high—certainly well above the 5x purchase price of the network-connected Wintel PC. If this cost of ownership cannot be spread across a large number of switches that can be uniformly configured and remotely managed (and optimally, wringing the most use from high-priced WAN bandwidth), ATM switches will have a hard time leaving the lab except to support the most exotic or profitable of applications.

As an industry, data communications has lurched from complexity to complexity: from SNA's virtual telecommunications access method/network control program (VTAM/NCP) in the 1970s to the Department of Defense's TCP/IP, Novell Netware's sequenced packet exchange/internetwork packet exchange (SPX/IPX), and Cisco's internetworking operating system (IOS) more recently. Each have proven to be a valuable technology—and a support nightmare. Arguably, ATM is an order of magnitude more complex than any of these worthy predecessors. Unless user-friendly, intelligent, self-adaptive, and remotely manageable software emerges, ATM is likely to have the highest ratio between "box cost" and support cost of any technology on today's horizon.

4.3 MANAGEMENT

Switch management can be likened to entertaining. With a large, formal dinner party, timing, synchronization, service type, prior arrangements, and resource provisioning is complex and essential to success. In contrast, at an informal buffet-style dinner (say, a backyard barbecue), virtually all requirements, short of minimal food and drink and a place to get out of the weather, are relaxed. Indeed, the guests need not show up at all—nor are there any guarantees that the food or drink will not run out!

Where an Ethernet LAN can be likened to the backyard barbecue, the ATM environment, particularly over the wide area and where vendor-heterogeneous equipment is involved, is a State dinner. Symptomatic of this complexity is that although the International Telecommunications Union (ITU) (formerly, the Consultative Committee for International Telecommunications & Telephony, CCITT) completed the cell-based, multimedia ATM standard in 1988, and the industry group ATM Forum has aggressively worked since 1991 to add flesh to the skeleton, in 1996 ATM switches made by different manufacturers still could not interoperate without sacrificing many of ATM's most attractive features.

The apparent solution to this situation, application-specific standard subsetting, has already been suggested. By example, in the 1970s the X.25 standard for packet switching spawned a large number of public and private data networks. Many smaller suppliers did a brisk business in just providing just a piece of the action—the CCITT X.29-conforming packet assemblers/disassemblers (PADs) or CCITT X.3 terminal control features. Of the number who fielded switches to do the actual packet switching, few implemented more than a subset of the large specification. In the early days, they verified their product interoperability by demonstrating interoperability with the largest of the public data networks (PDNs). Only much later did the U.S. government's National Institute for Standards and Technology (NIST) define an elaborate certification test suite.

As noted, management in ATM is at several levels. First, there is the operations administration and management monitoring (OAM, or OA&M) level, which is basically the monitoring of virtual circuits. Second, there is the evolving state of network management, the management of various—and oftimes hetereogeneous manufacturer—network elements.

The OAM monitoring is often referred to as the *flow reference* or *management plane* architecture. It consists of a number of functions performed by specialized OAM cells. These functions define the aspects of an ATM point-to-point virtual circuit (VC) that can be monitored and controlled and consist of the following:

- Fault and performance management (operations);
- Addressing, data collection, and usage monitoring (administration);
- Analysis, diagnosis, and repair of network faults (maintenance).

The flow reference model divides a VC into five distinct layers, labeled F1 through F5. It also defines the flows of ATM cells through these layers. The F levels are as follows.

The F1 level defines the flow of cells at the lowest physical layer of the ATM stack, the SONET (synchronous optical network) section layer (also known as the regeneration section level). A typical transmission path for cells at F1 would be through a SONET repeater in a WAN.

The F2 level defines the flow of cells at the SONET line layer (also called the digital section level). An example of an F2 function is the transmission of cells between two lightwave terminal equipment devices in a SONET network.

The F3 level partially defines the flow between a virtual path (VP) and a VC. In a large ATM network, a VC typically joins a VP, traverses it, and then splits out into a separate VC again. Traffic is forwarded from the

VC to the VP and back to a VC again via a cell relaying function (CRF). F3 defines the flows between the VC and CRF and between two CRFs.

The F4 level completes the definition of the traffic flow between a VP and a VC. F4 describes the transmission of cells from an end station, across a VC, through a CRF, and onto a VP. F4 stops at the second CRF (also known as the VP CRF).

The F5 level completes the definition of the traffic flow from one ATM end station to another. The flow goes from VC to CRF to VP and VP CRF, then to VC again, and finally to the destination.

In network management, today's interim specification is called the interim local management interface (ILMI). It functions between an end user and a public or private network and between a public network and a private network. (These relationships are depicted in Figure 4.1.) Defined by the ATM Forum, it is part of the user-to-network interface (UNI) 3.1 and is based on the simple network management protocol (SNMP). ILMI requires each UNI 3.0 end station or ATM network to implement a UNI management entity (UME). The UME functions as an SNMP agent that maintains network and connection information specified in an ILMI

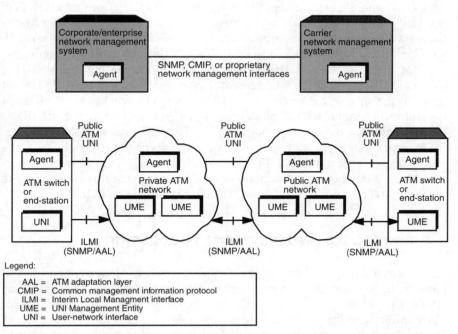

Figure 4.1 ATM network management today. (*Source:* Alexander, P., and K. Carpenter, "ATM Net Management: A Status Report," *Data Communications*, Sept. 1995, p. 111. © 1995 The McGraw-Hill Companies. All rights reserved.)

management information base (MIB). Based on a limited subset of SNMP capabilities, it responds to requests from SNMP management applications. Standard SNMP management frameworks can use AAL 3/4 or AAL 5 to encapsulate SNMP commands in ATM protocol data units (PDUs).

In the future, the ATM Forum promises a much more comprehensive network management facility based on a five-layer management model (see Figure 4.2). There will be five key management interfaces, management interface (M) M1 through M5. M1 concerns the management of ATM end devices and M2 the management of private ATM networks or switches. M1 and M2 are based on the IETF's SNMP-based MIB II and ATM management objects (AToM) in IETF's RFC 1695. M3, the customer network management (CNM) interface, defines the interface between customer and carrier management systems. M4 provides network and element management level (NML, EML) views into the carrier's network management system and public ATM network. M5 is the management interface between a carrier's own network management systems.

Today's commercially available ATM management tools reflect the two levels: OAM and network element. At the OAM level is a growing base of diagnostic tools where the cost continuum runs from line testers at the low end, through LAN analyzers, and up to protocol testers. At the network element level are numerous SNMP-based systems and a much smaller number of CMIP monitors and/or VT100/200 craft terminals. Except for high-end, large switches targeted for the carrier environment, the vast majority have written their switch management software for one or more of the three most popular SNMP-based UNIX workstation/net manager environments: HP's OpenView, Sun's Sun Net Manager, or IBM's NetView 6000. Conceivably, particularly if SNMP's capabilities continue to be improved, these may prove adequate to manage private, homogeneous manufacturer networks. But in cases where a private network incorporates heterogeneous manufacturer equipment or employs one or more public network "clouds," one cannot help but be pessimistic.

4.4   PROTOCOLS AND PROTOCOL PROCESSING

Rarely has a metaphor been as apt as diplomacy is to electronic communications. In a peace treaty, for instance, one encounters careful identification of the parties, time synchronization, event sequencing, provisions for verification, careful modularization of complex issues, official languages, treaty management provisions, attached technical materials like maps of terrain and minefields, and provisions for arbitration and modification. A data communications specialist would feel right at home.

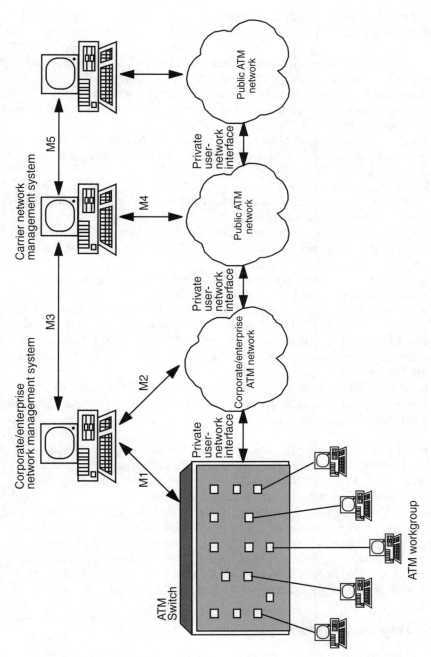

Figure 4.2 Future plan for ATM network management. (*Source:* Alexander, P., and K. Carpenter, "ATM Net Management: A Status Report," *Data Communications*, Sept. 1995, p. 114. © 1995 The McGraw Hill Companies. All rights reserved.)

Where do protocols (present and future) stand vis-à-vis high-speed networks? Already, the following contradicting points have been made:

- Many of today's most popular protocols (TCP/IP, a prime example) are "heavyweight" affairs, designed to overcome the deficiencies of noisy analog lines and impose heavy protocol overheads; this overhead is computed in terms of administrative bits, bytes, or octets per corresponding data units.
- Many of today's protocol stacks (DoD, ISO, SNA) incorporate logical redundancy where the data integrity is checked multiple times at multiple layers.
- Broadband network designers point out that error rates over fiber-optic links are rare, almost astronomical events, and current protocols impose needless overhead.
- Hard-nosed network managers point out that few networks are fiber optic end-to-end and that with big bandwidth pipes, infrequent errors are a frequent occurrence.
- Low-level protocols (ISO layer 4 and down) do not add value to the application from the user's view; users cannot tell whether their application is running in AppleTalk, SNA or TCP/IP; the raison d'être of communications is to deliver a useful product or service to users.
- Deadline-driven implementers advise, "Don't worry about the protocols; we'll encapsulate it and no one will be able to tell the difference."
- Protocol implementers say, "We'd rather not do the extra protocol processing but hey, if the customer wants it, we'll do it in parallel on whiz-bang RISC processors."

Finally, it has been suggested that two tendencies will be apparent regarding protocols and protocol processing on high-speed broadband networks. First, most current applications will be run in encapsulated mode for addressing and other transition issues. As already observed, the efficiency of these applications is unlikely to exceed, and may fall well below of, the asynchronous ASCII of the 1960s and 1970s. Second, for many new applications, particularly compressed video, high efficiency is essential to attractive pricing. A new VBR-RT standard would be one candidate; if the ATM Forum is perceived as unresponsive, it is likely that one or more "lightweight" protocols will emerge from that industry.

4.5 CLASS OF SERVICE

Most Americans are aware of class of service from airline seating, cruise ships or European trains, where with the latter whole cars are often marked in large, easy to read numbers showing "1," "2," or worse. In communications, class of service is typically encountered in the voice private branch exchange (PBX) environment. There, each user has a class of service associated with a station location (fixed) or identification number (mobile). This allows the switch administrator to enforce company policies by restricting certain employees for outside calls, or long distance access, or categorically preventing employee off-hour access to the company long distance resources.

With multimedia, use of ATM class of service becomes much more complex, particularly with public or shared facilities. With public facilities, class of service is linked to vastly different tariff rates; with shared facilities, as in a company setting, class of service may be translated into denial of service or probabilistic service, as determined by the corporate pecking order.

How might this public or shared access work in practice? The best metaphor reflective of the complexity involved is found in today's football stadia [1]. At the top of the heap are the sky boxes, leased on a season by season basis by the owner, chums of the owner, and corporate moguls. They represent large, permanently allocated bandwidth at high prices. They remain allocated even if empty. Next are the season ticket owners who hold seats of varying desirability (bandwidth). They can usually turn their tickets in for same price resale (the white market) or attempt to resell them for above the face value (the black market). Whether one wants to admit it or not, premiums for one-time access, high installation charges, high termination fees, minimum service periods, and similar mechanisms all have a black market aspect to them. Note thus far the discussion has focused on allocated, or, to use the old-fashioned term, "nailed up" bandwidth. The bandwidth has been allocated even though it may change hands through a secondary market mechanism.

Essential to the efficiency of an ATM switch is a sizable proportion of unallocated bandwidth, or, to continue the football metaphor, seats for sale on a game by game basis. These seats go on sale by a "fair" mechanism like a lottery or by endless, all-night lines at the ticket office. Oftentimes, demand for seats can exceed supply and ticket seekers are turned away or queue at the last minute for seats from the white market. A final category, standing room, can be likened to a low-priority in-switch queue, where, similar to near the end of the game, a lopsided contest, or inclement weather, normal seatholders abandon their seats. Overall, the objective of the football owner and public or shared ATM switch administrator is the same: Based on class of service, implement a rational system that optimizes capacity and benefit.

Commercial vendors have gone every which way with these basic classes. Some data switches offer only a single class, such as UBR. Some that offer only UBR employ proprietary means to deliver what is effectively an ABR service. Others mix and match to boast as many as 32 separate classes of service. Vendor-created QoS descriptors such as "SNA" and "IPX" have appeared. Uncritical use of these combinations may impede heterogeneous switch interoperability.

One reason for these vendor "shorthands" is that the configuration side of ATM's QoS is complex. UNI version 4.0 establishes explicit QoS parameters. There are three delay parameters (peak-to-peak cell delay variation [CDV], maximum cell transfer delay [Max CTD], mean cell transfer delay [Mean CDV]) and one dependability parameter, the cell loss ratio [CLR]. CDV, Max CTD, and Mean CDV, the delay parameters, are combined into a dynamic, additive index with their expected values contained in UNI and P-NNI signaling requests. CLR, the dependability parameter, is a configured link-and-node attribute, which local connection admission control (CAC) algorithms must meet.

4.6 SECURITY

One of the fundamental paradoxes of networks and networking is that their utility is inversely proportional to their security. To this end, it is instructive to contrast the two poles of today's security continuum, a commercial systems network architecture (SNA) network and the Internet.

In a traditional corporate SNA network, the mainframes are housed in at least two different physical complexes, often in different states (different power grids, local exchanges), and are mutually supporting. Since the 1960s and the latest resurgence of corporate vandalism/terrorism, rarely are the computer centers, often located in anonymous industrial parks, identified by signs. Even when one calls the facility, the answering party identifies only the extension number. Employees are background checked, badged by physical area on a need-to-know basis, and often, in the financial industry, bonded. Card key systems, sensor and alarm systems, fencing and lighting, video monitoring, even armed guards at the site, are not unusual. Such "fortress" features reduce insurance rates. The telecommunications side consists of leased lines connecting the alternative site and remote controllers that concentrated the terminal traffic (usually IBM PCs performing 3270 emulation). Communications flows are straightforward: the logical paths and physical paths are, for normal operation, the same. In its purest form, no dial-in service is available. In the financial industry, some applications, such as money transfers, are encrypted using the data encryption standard. The terminals, in their turn, are in office environments that also require identification cards and/or cardkeys. Each employee has an individual logon/password and often a secondary barrier such as project number and/or database password that is verified by an access control system such as RACF or ACF2. Rigid check-out procedures are followed whereby passwords and cardkey access are invalidated upon termination or retirement. If "run by the book," such a corporate computer/communications complex makes a difficult target.

In the Internet, one has millions of users, perhaps tens of millions, worldwide. Virtually every imaginable group is represented, including dedicated hackers and a variety of groups with antisocial proclivities. Targets abound, as government agencies such as the U.S. Departments of Energy and Defense, the Federal Bureau of Investigation, the Secret Service, and the Central Intelligence Agency all have accounts. User skill levels range from rank beginner to expert. Many organizations, particularly universities, offer free public access along with all sorts of specialized databases; commercial services offer a plethora of services. Net "surfing," what the security folks would characterize as "doorknob rattling," is institutionalized, even automated with publicly available "knowbot" software. What security exists is distributed and ranges from the primitive to the elaborate and expensive—not necessarily the same as effective. Some sites have installed "firewalled" access to systems and some have systems so lazily administered that it amounts to none. The Internet, as a number of observers have remarked, reflects the society that uses it.

In comparing these polar opposites, one sees that the SNA network is quite secure but in today's highly mobile, freewheeling, and collaborative business environment, perhaps of limited utility—or an actual business detriment. By contrast, the Internet is ubiquitous, greatly aids human information exchange but its insecurity can invite unwelcome surprise. These unwanted intrusions can range from simple embarrassments (unauthorized use or unsuspected modification of network resources) to out and out disaster (virus infection, fraud, or theft).

Finally, security systems and their costs can be directly compared with the penal system. Parole and work release programs have low-cost but poorly run programs that often endanger the public. Maximum security prisons are very expensive (in the neighborhood of $45,000 per inmate yearly in the United States) and only very rarely endanger the public. Current computer security technology offers no "silver bullet." Other than their expense, which often occasions more vigilant oversight, there is no reason to believe that broadband networks will be any more or less secure than their predecessors. In a dynamic that some have compared to the 100-year relationship of armor to the armor-piercing shell, stealing data or tapping voice conversations, even when packaged in cells that zip by in microseconds, will not be beyond the reach of either the nefarious and/or the technologically astute.

4.7 FAULT TOLERANCE

Higher speed networks are akin to super highways in that the number of accidents per passenger mile may be lower, but when they occur, more people are likely to be affected. To continue the automobile analogy, most transient hardware or software errors simply cause switch performance degradation, similar to a car that runs but is out of tune. With high-speed switches, the degradation may be subtle to the point that it is not perceived; this is why subsequent software releases for a switch almost always improve performance. Other times the bugs or transients are not successfully cleared, localized, or "ridden through" and the switch sheds traffic or self-administers a shut down-and-reboot cycle. This is akin to a car on a superhighway losing power and having to pull to the shoulder. The most destructive of the switch failures, which can be likened to a freeway pileup, is when a bug in the switch-to-switch control software has one failing switch propagating failure modes over the network [2]. In this case, the entire network may lock up in the manner of horrendous pileups so dense that wreckers and ambulances cannot access the accident scene and the police are forced to close the highway.

Aside from noting hardware transients and software bugs that often appear and disappear, what causes these errors? In switches that have been carefully engineered and that have undergone an extended beta testing, subsequent errors are often characterized as "rare events." (See Figure 4.3.) That is, some combination of feature use or unusual traffic occurs, and the switch does not handle it in a routine manner. An order of magnitude more rare, and potentially lurking in every complex switch software suite regardless of age is the joint occurrence of rare events, essentially the interaction of rare events.

It is worth noting that oftentimes the switch failures are abetted or tripped by exogenous events, often involving well-meaning human intervention. A failure of an AT&T Manhattan switching office was initiated by employees responding to the local electric company's call to minimize power consumption; this action led to a series of mishaps that left the facility without power. The power failure in the switching office then led to the shutting down of air traffic control radars in much of northeastern United States [3]. In a well-known case, the history of false fire alarms at a large, unmanned Chicago area central office (and AT&T long distance point of presence, POP) led network controllers to shut off the alarm

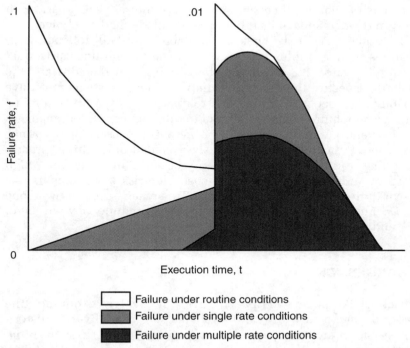

Failure under routine conditions
Failure under single rate conditions
Failure under multiple rate conditions

Note: depicted above are progressively smaller tail areas (representing the frequencies of rare error conditions) under one portion of a normal curve.

Figure 4.3 The progression of failure causes. (*Source:* Abouelnaga, A., et al., *Report of the Voice Switching and Control System (VSCS) Independent Fault Tolerance Analysis Team (VIFTAT)*, Martin Marietta, Jan. 1993, pp. 1–6.)

rather than call for help. By the time the real fire was confirmed and reported locally, the entire facility was lost, causing a massive service outage that lasted weeks [4].

Why do these accidents occur and reoccur? It would seem that mankind (and womankind) is destined—if not impelled—to create structures that overstep limits of materials, design, or control. In medieval Europe, cathedral spires were continually thrust up higher and higher in an effort to reach, in a form of homage, to God. As the materials and workmanship were variable and the side forces imperfectly understood, the spires often fell down, were rebuilt, and often fell down again. In our commercially impelled modern society this hubris continues and thrives. Software size and complexity grows geometrically; network control facilities, crammed with computerized monitoring equipment and slim staffs of experts, remotely control thousands of automated teller machines, hundreds of

PBXs, dozens of central office switches. Five hundred years apart, each group, given the demands of its religion, concluded it had no choice.

All of today's ATM switches are available with fault-tolerant hardware features such as duplex processor cards, redundant line cards, dual power supplies, and "1 + 1" circuit provisioning. But the software, the most volatile aspect of the system, is simplex. (Some systems may have the capability to reboot to a previous release.) "*N*-way" programming, where the same requirements are written in different computer languages to run on several parallel (and usually different) processors, or "voting" systems where the code is synchronously executed on multiple processors and the results compared, remain—because of their very high cost—largely limited to real-time, high-cost, high-risk environments like "fly-by-wire" aircraft control. For today's ATM switches, if a software bug like the DSC bug escapes detection, the hardware redundancy will be of no avail, and the network will fail.

4.8 CHARGEBACK

Chargeback is like paying the paper carrier, it involves a product (the newspaper or data bandwidth) and a reliable service (delivery). Historically, most packet switching systems have charged for (1) system initiation and/or termination, (2) number of packets transferred, (3) network connect time, (4) and a two- or three-level surcharge based on time of day and day of week. ATM, as the latest and most sophisticated incarnation of packet switching, offers a bewildering set of options with reservations (emulation, guaranteed, available, etc.), physical interfaces (TAXI, DS-1, DS-3, V35, etc.), protocol conversions (adaptation layers 1, 3/5, etc.) and services (LAN emulation). The earliest fielded ATM systems have targeted the LAN/laboratory market where most of the circuits are internal to the organization and chargeback schemes are generally perceived as more trouble than they are worth. But once ATM switches are used in multiapplication environments where high-price switched or leased circuits are involved, chargeback becomes a political, if not financial, necessity.

Packet switching is logical multiplexing and should, given several data streams, "pack" the bits to be transferred more economically into a given pipe than if the several applications employed several separate pipes used intermittently. This is the so-called mux effect or mux savings. Currently, commercial frame relay vendors (perhaps the closest comparable service to ATM) generally price their services about 25% under what the comparable pipe (DS-1, DS-3) would cost monthly.

One of the things that makes chargeback so much fun is that it brings out Robin Hood behavior in a sizable number of users. Just like tax systems breed tax accountants, armor plate begets armor-piercing shells, and chargeback algorithms attempt to mold human behavior and inevitably encourage gaming. For example, today's popular online services try to encourage use by keeping their rate structure simple and understandable and the charges for connect time low. Savvy users who spy flat-rate access charges often upgrade their modems so as to wring more action—pass more bits—through the network, which can increase the vendor's costs. To cite just one possibility in the ATM case, any user that finds out that the ABR and UBR services are, in fact, equivalent in performance, will immediately gravitate to the latter, cheaper service. With ATM's complex class-of-service hierarchy, the service vendor is caught in several dilemmas relative to making a profit. If the vendor makes the rate structure too complex (such as today's long distance-voice market), the buyer will demand 30-day trials and other expensive, performance-based comparisons. Further, the vendor faces constant attack from other vendors who will attempt to steal customers through understandable (often flat-rate) pricing schemes. Keeping customers will require fighting back and increased advertising budgets. Bargain-basement or flat-rate vendors face even another risk, that they themselves will be transformed into commodity carriers by clever resellers. As suggested, chargeback and chargeback schemes guarantee an exciting life, but not necessarily a happy one.

4.9 SYNCHRONIZATION AND TIMING

Synchronization uses shared signals to get the network elements to work together as a whole. A nice analogy is a military parade. A small unit, such as a platoon, can time-keep and synchronize very well by itself, particularly if the parade path is not too noisy. A larger unit, such as a battalion, extends for several city blocks and necessitates a military band to stay in step.

The conventional approach to synchronization was developed by the Bell System in order to implement long distance telephone service. It consisted of a clock hierarchy of progressively more accurate (and more expensive) system clocks whose signals were propagated downward via the network to the network elements. Even in this tidy setup there were problems with clock "wander" (the effect of cumulating inaccuracies) and improperly compensated for delays due to subtle differences in signal latency. So, similar to the military units' platoon guide, a sort of small unit reference point for local alignment, other engineering adjustments were necessary. Proper digital synchronization requires certain binary bit patterns ("flags") to arrive within a certain timing window. When they do not, special buffers are used to accommodate the overflows (or underflows) and the timing is locally adjusted. Today's "robbed bit," 24 DS-0 channel T-1 service works this way.

The communications environment has changed considerably since the AT&T divestiture. There are many timing sources today, including the use of satellite references. Too, the traffic types have changed. Where voice transmission is remarkably resistant to the effects of errors but is intolerant of delays, data traffic requires mostly error-free transmission, but many data types can be delayed without serious effect. Many of today's local data environments resemble the "self-clocking" platoon; one box provides the clock and the others stay in step.

Obviously, high-speed networks tend to exacerbate timing and synchronization problems. Clocks need to be more accurate (and more expensive). Buffers need to be more copious. Transmission timing faults affect more circuits and more users (whether user-visible or not) and can cause efficiency-robbing retransmissions. All of which has accelerated the deployment of the fiber-based synchronous optical network (SONET). Often deployed as topographical rings for increased fault-tolerance, SONET transmission facilities enable the economic operations and management of very large, very accurate bandwidth bundles [5]. In WAN environments, both synchronous transfer mode (STM) and asynchronous transfer mode (ATM)-based services ride on top of SONET facilities. Here, a nice distinction can be made between the STM and ATM approaches. In STM, the packets or cells carrying applications data have to be rigidly aligned with the underlying synchronous frames; in ATM, the cells carrying applications data can be independent of (or *asynchronous* to) the underlying synchronous frames.

4.10 SUMMARY

In this second chapter on the technical challenges of high-speed switching, another set—compression, cost, management, protocols and protocol processing, class of service, security, fault tolerance, chargeback and synchronization and timing—of well-known problems or, more optimistically, opportunities for heroic switch designs, have been overviewed. Such a recounting is intended to remind the reader that these challenges are recursive—they do not go away with new technology and, more ominously, are often exacerbated by higher and higher speeds.

End Notes

[1]   Inoue, Y., et al., "Granulated Broadband Network Applicable to B-ISDN and PSTN Services," *IEEE Journal on Selected Areas in Communications*, Vol. 10, No. 9, Dec. 1992, pp. 11474–1488.

[2] Andrews, E., "String of Phone Failures Reveals Computer Systems' Vulnerability," *New York Times*, July 3, 1991.

[3] Bamford, D. L., L. Monticone, and T. W. Kennedy, "New York ARTCC Communications Outage," Sept. 17, 1991, MITRE Briefing F85-B-245.

[4] National Communications System, "May 8, 1988 Hinsdale, Illinois Telecommunications Outage," Aug. 2, 1988. Cited in McDonald, J. C., "Public Network Integrity—Avoiding a Crisis in Trust," *IEEE Journal on Selected Areas in Communications*, Vol. 12, No. 1, Jan. 1994, pp. 5–12.

[5] The "mother of all SONET rings" is currently being laid around the continent of Africa by a consortium that includes AT&T.

Commercial ATM Products 5

One of the most entertaining aspects of technology is how the inventors—or, in the case of ATM, the specifiers—conceived the technology would be used, and how the vendors and users conspired to use it. In the case of ATM, what has been called the "Magic of the Market" has been especially active. A number of these aspects are discussed below.

5.1 PRESTANDARD

If the time-honored way to market share and niche dominance is being the first with the mostest, what does one do when a standard is clearly needed and yet the standard is not there? The answer is the "prestandard," or the proprietary implementation that makes the product useful to users and represents the company's best guess at what the standard implementation will look like.

Both FORE Corporation, a market leader at the low end, and StrataCom (now Cisco), at the network edge, have traveled the prestandard route. FORE satisfied demands for high-speed LAN interfaces with a 140-Mbps version of the TAXI interface that was later rejected by the ATM Forum (the 100-Mbps version was approved). Also, as the first users struggled with trying to communicate with only PVCs, FORE's first implementation of SVCs was "prestandard." In the case of StrataCom, they already had a successful cell switch for data (most of the commercial frame relay networks used it), so they backed it into the ATM standard. As WAN switches have to talk with other WAN switches to be useful, their homogeneous WAN ATM networks relied on the same proprietary congestion control that had proven successful in the frame relay application area.

Neither vendor has any illusion about the longer term efficacy of proprietary solutions in the ATM marketplace. Heterogeneous switch connectivity is clearly the overriding mandate, even among the early implementers. As soon as the ATM Forum produced the appropriate standard, the vendors implemented it and then worked hard to demonstrate interoperability with other vendors products.

5.2 NETWORK INTERFACE CARDS (NICs)

NICs are hardware and software that connect the end system, which may or may not be the end user, with the communications network. At the very least—and they can do a great deal more—they communicate an end-system address and perform the segmentation and reassembly (SAR) processing. Ideally, but dependent on an applications interface, they could provide the user with a means to interact with the classes of service offered by the ATM network, rather like a TV remote allows the TV watcher to select programs and fiddle with the sound. Unfortunately, the current NICs are without applications interfaces and the user is essentially operating within a series of presets established by the switch's systems administrator. Rather than the bandwidth on demand, current practice more greatly resembles an army chow line.

Nonetheless, NICs play an essential role in the ATM switching milieu. They are the most numerous network element and so their cost is important. It is nice if they are cheap, under $200 or $100, or some other marketing "sweet spot." As noted with Ethernet NICs, there is a clear and established inverse relation with price and popularity. In no circumstances should a NIC exceed the cost of the workstation or device they are attached to. Although that seems self-evident, it maybe a close call if they become part of onboard video servers on airplanes, where the display screen is necessarily small.

There are a number of observations one can make today regarding the "first generation" NICs (Table 5.1) [1]. First, as there was no installed based of ATM switches, the NIC market had to be "jump-started" by switch vendors like FORE who could not sell their switching products without NICs. With the relatively low volumes involved and the multiplicity of bus types, this tended to make prices high, over $1,000 per unit. Second, ATM hardly has the high-speed LAN market to itself, with competition from F/CDDI and several varieties of fast Ethernet. Although it would appear that ATM scalability and speed give it a clear edge, the competition and uncertainty keep volumes low and prices relatively high. Third, and more troubling, are the questions regarding noninteroperability and/or limited functionality of the first-generation NICs. Although small labs can readily sample and discard technologies, businesses, which buy in quantity and expect reasonable equipment lifetimes, are likely to hold back. Many of today's NICs will only work with one or a small number of ATM switches. Likewise, many of today's NICs are enabled for a single application area, such as moving data using AAL 5. These limitations must be overcome before ATM takes its place as a serious commercial communications technology.

5.3 PLAIN CREATIVE MARKETING

One of the earliest and most confusing ATM marketing ploys has involved combining service and QoS types to produce new acronyms to spring on the unwary. For example, noting SNA's well-known performance target of small timing variations per transaction and that session disruptions that can be triggered by timeouts, a prerolled vendor QoS called "SNA" has been born. Similarly, the statistical multiplexing game is being driven over the cliff. One starts with constant bit rate (CBR); CBR is like obeying the 55-mph speed limit, if you only do 55 mph, you do not need to worry about speed traps. Intuitively, one can comprehend that even with huge port buffers on an ATM switch, only so much traffic can be expeditiously passed over fixed amounts of physical bandwidth. This is the basis of the variable bit rate (VBR) service. To this is added the recently completed available bit rate (ABR) specification, which can be likened to airline overbooking, where a statistically small number of times will result in ticketed passengers being denied a seat. Prior to ABR, vendors hawked "unspecified bit rate" (UBR), which can be likened to showing up at the airport without a ticket. All this would be huge fun were it not that dissimilar switches will eventually have to agree on connection parameters to communicate and that "roll your own" QoS categories may lead to mis-specified or failed connections.

5.4 THE MANAGEMENT SHORTFALL

In the rush to demonstrate ATM-enabled, high-bandwidth applications, the very last consideration was providing for test and measurement and network management support. One account of a heterogeneous vendor test site allowed that they had one piece of supervisory equipment for every piece of working communications equipment—a number far exceeding the workstations on which the application was running. Network test equipment can be expensive as well; not only is ATM complex, but the buffers, memory, and disk capacity to capture and store even a few seconds of high-bandwidth transmission is significant. For example, HP's basic ATM analyzer, hosted on a UNIX workstation, starts around

Table 5.1
First-Generation NICs

Vendor	Product	Platform	Interface speed(bit/sec)/cabling requirements	Windows 95	Windows NT	Netware 3.X	Netware 4.X	SunOS and Solaris	Other	Price
		Bus type and hardware				Operating systems				
Adaptec, Inc. (408) 945-8600	ANA-5940 and ANA-5930	PCI bus-based PCs	155M/fiber with SC connectors or UTP	×	×	×	×		MacOS	Fiber: $595; UTP: $495
	ANA-5910 EL and DX	PCI bus-based PCs	25M/UTP	×	×	×	×		MacOS	$199–$349
	ANA-5240 and ANA-5230	Sbus-based Sun SPARC workstations	155M/fiber with SC connectors or UTP					×		Fiber: $995; UTP: $895
	ANA 5210	Sbus-based Sun SPARC workstations	25M/UTP					×		$349
Cabletron Systems, Inc. (603) 337-9400	NBA-200 series	NuBus-based systems	155M/fiber with SC or ST connectors or UTP						MacOS	$1,940
	SBA-100 and SBA-200 series	Sbus-based Sun SPARC workstations	100M or 155M/fiber with STor SC connectors; UTP for 155M model					×		$1,075
	GIA-100 and GIA-200 series	GIO bus-based SGI workstations	100M or 155M/fiber with STor SC connectors; UTP for 155M model						Irix Indigo and Indy workstations	$1,075
	HPA-200 series	EISA bus-based HP 9000/7xx workstations	100M or 155M/fiber with STor SC connectors; UTP for 155M model						HP-UX 9.X	$1,075
	ESA-200 and ESA-200PC series	EISA bus-based PCs or SGI Extremeworkstations	100M or 155M/fiber with STor SC connectors; UTP for 155M model		×	×			SGI Irix 5.3	$1,075
	MCA-200 series	Micro Channel bus-based IBM RS/6000 workstations	100M or 155M/fiber with STor SC connectors; UTP for 155M model						AIX 3.2X	$1,400
	TCA-100	Turbochannel bus-based DEC workstations	100M/fiber with ST Connector						Ultrix	$1,075

(1) Price is for OEMs and resellers. (NIC: network interface card; SGI: Silicon Graphics, Inc.; UTP: unshielded twisted pair.)
Source: Mier, E., "It's Arrived," *Network World*, September 16, 1996, pp. 91–100. © 1996 Network World Inc., Framingham, MA 01701. Reprinted with permission.

Table 5.1 (Continued)

Company	Product	Platform	Media	OS	Price
Connectware, Inc. (214) 907-1093	CELLerity ATM 1110Sbus NIC	Sbus-based Sun SPARC workstations	155M/fiber with SC connectors or UTP		From $699
	CELLerity ATM 1110PCI NIC	PCI bus-based PCs withPentium processor	25M or 155M/UTP for 25M;fiber or UTP for 155M	SCO Unix	From $495
	CELLerity ATM 1110EISA NIC	EISA bus-based PCs and HP-UX workstations	25M or 155M/UTP for 25M;fiber or UTP for 155M		From $595
Digital Equipment, Corp. (800) 457-8211	ATMworks 350	PCI bus-based PCs	155M/fiber with SC connectors	Unix	$1,995
	ATMworks 950L	Sbus-based Sun SPARC workstations	155M/fiber with SC connectors or UTP		Fiber: $1,395; UTP: $1,295
	ATMworks 350L	PCI bus-based PCs	155M/fiber with SC connectors or UTP		Fiber: $1,095; UTP: $995
	ATMworks 750	Turbochannel bus-based workstations with DEC Alpha processor	155M/fiber with SC connectors	Unix	$2,495
Efficient Networks, Inc. (214) 991-3884	ENI-155p	PCI bus-based PCs	155M/fiber with SC connectors or UTP		Less than $1,000 (1)
	ENI-155 series	Sbus-based Sun SPARC workstations	155M/fiber with SC connectors or UTP		Less than $1,000 (1)
	ENI-100s	Sbus-based Sun SPARC workstations	100M/fiber with SC connectors		Less than $1,000 (1)
	ENI-155e	EISA bus-based PCs and PowerPCs	155M/fiber with SC connectors or UTP		Less than $1,000 (1)
	ENI-25p	PCI bus-based PCs with DEC Alpha, Pentium or PowerPC processor	25M/UTP	Windows and Netware Client 32 planned	$249
	ENI-100e	EISA bus-based PCs and PowerPCs	100M/fiber with SC connectors		Less than $1,000 (1)

(1) Price is for OEMs and resellers. (NIC: network interface card; SGI: Silicon Graphics, Inc.; UTP: unshielded twisted pair.)

Table 5.1 (Continued)

Vendor	Product	Bus type and hardware / Platform	Interface speed(bit/sec)/cabling requirements	Operating systems						Price
				Windows 95	Windows NT	Netware 3.X	Netware 4.X	SunOS and Solaris	Other	
Fore Systems, Inc. (888) 404-0444	HPA-200 series	EISA bus-based HP workstations	155M/fiber with ST or SC connectors or UTP						HP-UX	$995–$1,195
	SBA-200 series	Sbus-based Sun SPARC workstations	155M/fiber with ST or SC connectors or UTP					x		$995–$1,195
	GIA-200 series	GIO bus-based SGI workstations	155M/fiber with ST or SC connectors or UTP						Irix	$995–$1,195
	ESA-200E and ESA-200EPC series	EISA bus-based PCs and SGI workstations	155M/fiber with ST or SC connectors or UTP		x	x			SGI Irix	$995–$1,195
	NBA-200 series	NuBus-based systems	155M/fiber with ST or SC connectors or UTP						MacOS	$995–$1,195
	MCA-200 series	Micro Channel bus-based IBM RS/6000 workstations	155M/fiber with ST or SC connectors or UTP						AIX	$1,295–$1,495
	PCA-200 series	PCI bus-based PCs and Power Macintoshes	155M/fiber with ST or SC connectors or UTP		x	x			Apple OpenTransport 1.1	$995D$1,195
	VMA-200 series	VME bus-based SGI workstations	155M/fiber with ST or SC connectors or UTP						Irix	$1,995–$2,195
Interphase Corp. (214) 654-5000	5515	PCI bus-based PCs	155M/fiber with SC connectors or UTP	x	x	x	x		MacOS, OS/2, UnixWare, Solaris x86	$650
	5525	PCI bus-based PCs	25M/UTP	x	x	x	x		MacOS, OS/2, UnixWare, Solaris x86	$222

(1) Price is for OEMs and resellers. (NIC: network interface card; SGI: Silicon Graphics, Inc.; UTP: unshielded twisted pair.)

Table 5.1 (Continued)

Company	Model	Platform	Interface				OS	Price
Interphase Corp. (214) 654-5000	4615	Sbus-based workstations	155M/fiber with SC connectors or UTP	×				$795
	4815	PCI bus-based PCs with Pentium processor	155M/fiber with SC connectors or UTP	×	×	×	HP-UX, 9.X and 10.X, Irix	$895
	5215	VME bus-based workstations	155M/fiber with SC or ST connectors				Irix, SunOS, HP-RT	$2,795
	4915	GIO bus-based SGI workstations	155M/fiber with SC connectors or UTP				Irix	$995
Madge Networks, Inc. (800) 876-2343	Collage 25 Adapter	PCI bus-based PCs	25M/UTP	×	×		OS/2 Warp	$295
	Collage 155 Adapter	PCI bus-based PCs	155M/fiber with ST connectors or UTP	×	×		OS/2 Warp	Fiber: $1,295; UTP: $1,195
Newbridge Networks, Inc. (800) 343-3600	VIVID ATM NIC	EISA bus-based PCs, HP 9000 workstations, and SGI Challenge and Indigo2 workstations	155M/fiber with SC connectors	×	×		HP-UX, Irix	$1,267
	VIVID Sbus ATM NIC	Sbus-based Sun SPARC and Ultra workstations	155M/fiber with SC connectors			×		$1,267
	VIVID GIO ATM NIC	GIO bus-based SGI Indy Challenge servers	155M/fiber with SC connectors				Irix 5.X	$1,267
	VIVID PCI ATM NIC	PCI bus-based PCs, PowerPCs, and DEC AlphaServers	155M/fiber with SC connectors	×	×		DEC Unix	$1,267
Olicom USA, Inc. (214) 423-7560	RapidFire OC-615x	PCI bus-based PCs with Pentium processor	155M/fiber with SC connectors or UTP	×	×	×	OS/2 Warp	Fiber: $795; UTP: $695
SDL Communications, Inc. (508) 238-4490	RISCom/TCi card	PCI bus-based PCs with Pentium processor	155M/fiber with SC connectorsor UTP (T-1 and T-3 models available)	×			BSDi, LINUX	From $1,500

(1) Price is for OEMs and resellers. [NIC: Network interface card; SGI: Silicon Graphics, Inc.; UTP: unshielded twisted pair.]

Table 5.1 (Continued)

Vendor	Product	Platform / Bus type and hardware	Interface speed(bit/sec)/cabling requirements	Operating systems						Price
				Windows 95	Windows NT	Netware 3.X	Netware 4.X	SunOS and Solaris	Other	
Standard Microsystems Corp. (516) 273-3100	ATM Power 155 Sbus series	Sbus-based workstations	155M/fiber with ST connectors or UTP					x		Fiber: $1,295–$1,990 UTP: $1,149–$1,880
	ATM Power 155 PCI series	PCI bus-based PCs	155M/fiber with ST connectors or UTP	x	x	x	x			Fiber: $1,149–$1,880 UTP: $995–$1,750
	ATM Power155 EISA series	EISA bus-based PCs	155M/fiber with ST connectors or UTP	x	x	x	x			Fiber: $1,295–$1,990 UTP: $1,149–$1,880
Sun Microsystems, Inc. (415) 960-1300	SunATM	Sbus-based Sun workstations, including Media and Netra	155M or 622M/fiber or UTP for 155M;fiber for 622M						Solaris 2.4 and later	$995–$4,995
SysKonnect, Inc. (408) 437-3800	SK-7321 and SK-7341	EISA bus-based PCs	155M/fiber with SC connectors or UTP		x	x	x			Fiber: $1,495; UTP: $1,195
	SK-7521 and SK-7541	PCI bus-based PCs	155M/fiber with SC connectors or UTP		x	x	x			Fiber: $1,495; UTP: $1,195
	SK-7621 and SK-7641	Sbus-based Sun SPARCstations	155M/fiber with SC connectors or UTP					x		Fiber: $1,295; UTP: $1995

(1) Price is for OEMs and resellers. (NIC: network interface card; SGI: Silicon Graphics, Inc.; UTP: unshielded twisted pair

Table 5.1 (Continued)

						Price[1]
3Com Corp. (800) 638-3266	ATMLink Sbus-155 Fiber	Sbus-based workstations	155M/fiber with SC connectors		×	$1,495
	ATMLink PCI-155 Fiber	PCI bus-based PCs Pentium processor	155M/fiber with SC connectors	×	×	$1,495
ZeitNet, Inc. (408) 986-9100	ZN-1211 and ZN-1215	Sbus-based Sun workstations	155M/fiber with ST connectors or UTP		×	Fiber: $995; UTP: $895
	ZN-1221 and ZN-1225	PCI bus-based PCs	155M/fiber with ST connectors or UTP	× × ×		Fiber: $1,095; UTP: $995

(1) Price is for OEMs and resellers. (NIC: network interface card; SGI: Silicon Graphics, Inc.; UTP: unshielded twisted pair.)

$75,000. Finally, there is growing awareness that current ATM monitoring capabilities, most typically built on SNMP and RMON, may not be up to the challenge. Several consortia are attempting to move the ATM Forum towards greater capability. But for today, relative to the size and complexity of the management problem for low-cost remote monitoring capability or an integrated ability to monitor several manufacturers' equipment, the commercial products simply are not there.

Another aspect of the early lab emphasis is the lack of sophisticated billing and chargeback software. Most of the costing algorithms offered by the public ATM providers seem to be a combination of a bandwidth discount and amortized switch costs for "special needs" customers. As of today, their billing systems are not ready for a mob of customers with very different needs and pocketbooks, much less for badly behaved users.

5.5 YOUR MILEAGE WILL VARY

Where the ATM glamour has been multipoint viewing of MRI scans or interactively accessing huge, remote data repositories, more than a few users have sat down and (often with very inferior test equipment) actually tried to measure their throughput. Most of these dour types seem to have shared the same AAL 5, UBR, TCP/IP orientation and their results, before tuning, have been almost unbelievably discouraging. One European PTT is reported to have achieved zero throughput through an apparently optimally unoptimal cycle of transmit, overflow, discard, retransmit, and so forth. Many have reported only modest improvements in throughput over dedicated or switched Ethernet environments. Others have seen SAR processing either consume all their workstation's cycles, or, when unbalanced, cause one end to overflow and crash. And, invariably, when one views quality ATM-delivered video at trade shows it turns out to be CBR.

So what have we learned thus far? Perhaps first and foremost is that when one is dealing with many megabits per second, application tuning and the test equipment to carry it out are essential. With many ATM port buffers sized at a megabyte or less, and secondary buffer pools (if available) ultimately finite as well, both effective switch congestion control and the ability to quiesce the source are essential. Clearly, ATM planning is much more akin to the world of statistical multiplexers than Ethernet. Particularly when common carrier charges are involved, one must remember that "plug and play" and "bandwidth on demand" are mostly marketing fog.

A second observation is that by unhappy accident, ATM switches are first appearing as fast LANs or "top" devices in a collapsed backbone of LANs. Where in voice or video applications cell discards might cause pops and snaps, cell discards with computer data cause ruptured packets, and ruptured packets trigger retransmissions. Although several techniques—early packet discard (EPD) and trailing packet discard (TPD)—can mitigate the worst effects of TCP-initiated retransmissions, retransmission "storms" will very likely continue to plague WANs combining routers and ATM switches [2].

5.6 THE LEGACY OF THE LEGACY

If one aspect stands out about the current ATM products, it is the enormous efforts to accommodate the installed base of data communications equipment. Far from defining the multimedia future, most vendors are almost fixated on Ethernet LAN past, in many cases directly positioning ATM LANs against switched Ethernet or the several versions of fast Ethernet. An article of faith seems to be that as Ethernet NICs are the most numerous network component, these numerous NICs will dictate the technology of the LANs of the future. Although attractive as this theory may be to the spreadsheet set, it may not be correct; clearly, a number of high-speed multimedia applications are not likely to run, however compressed, satisfactorily on any Ethernet.

Nonetheless, this backward focus has been intense. The Forum's LAN emulation (LANE) standard has been an item of intense vendor interest and considerable technical exertion, as well as spawning several prestandard implementations. To this end, a number of ATM LAN products, switches, and NICs, only support AAL 5. Clearly, for today's ATM market, updating legacy LANs drives ATM "box sales."

Another area where legacy systems are torquing the ATM technology is in NICs. The aftermath of the PC wars has been too many bus types. In rough chronological order, there is the ISA (Advanced Technology, AT), MicroChannel Architecture (MCA), Extended ISA (EISA), and Peripheral Component Interconnect (PCI) buses, and this does not count Macintosh and the various workstation families. Of the PC products, probably only the new PCI bus is really suitable to attach to a high-speed LAN. Nonetheless, the vendors, in trying to accommodate this rich legacy of buses, are forced to produce many models of ATM NICs, which, in turn, tends to keep prices high.

5.7 PRESENTISM

Although concentrating on current products, one must not fall victim to a data communications cum workstation myopia. As noted earlier, if NIC costs fall to Ethernet levels, running various video applications is likely to be the overwhelming use of the ATM technology. It is easy to envision an entire ATM-enabled video "food chain:" cameras, transmission equipment, display equipment. Low-cost PCMCIA card (or smaller) implementations for consumer electronics opens up the possibility of ATM use in PBXs, cell phones, and personal digital assistants (PDAs) of many types where the ATM triumvirate of media multiplexing, class of service, and high speed are required.

5.8 TODAY'S SWITCHING VENDORS

Looking back over the communications products of the last three decades, cost/performance is probably the most important of the three attributes—the others being standards and time to market—that determine product success. The winners, such as Ethernet, routers, and TCP/IP had it; the losers, such as integrated systems digital networks (ISDN), token ring, and fiber distributed data interface (FDDI) did not.

Looking at ATM switches and switching, it has been previously argued that these multimedia-capable switches will be mostly used in overlay networks to enable high-bandwidth new applications—not in replacing older, less flexible in-place plant equipment. Further, many of these application areas are functionally very narrow, predominately high-speed delivery of a few kinds of data like video, X-rays, graphic images, and the like. The current batch of router manufacturers have demonstrated that very high packet processing rates (300,000 packets per second) are achievable by combining high-performance reduced instruction set computers (RISC) processors, logical parallelism, and application-specific logic-implemented firmware. Much faster routers wait in the wings and some will offer QoS-like features like RSVP. Most of these same manufacturers offer ATM products or router interfaces to ATM. The narrow focus on well-developed, low-end needs is key to popularity for ATM. It is not costly to produce small ATM switches, ATM interface cards, or even ATM computer channels that move data, typically in controlled environments, at OC-3 (155 Mbps) speeds. With these devices, there is no obligation to implement the full complexity of the ATM standard. Software costs are kept low, and, increasingly, software size over forecast unit sales determines product cost.

Consider the opposite case. A large edge switch must be able to serve many customers, input streams, and data and interface types. Further, the large edge switch is required to adjudicate between traffic types, avoid congestion, guarantee service levels, interface with other manufacturers' products, monitor switch performance, and, eventually, support the billing of users. To achieve this functionality requires implementing a great deal of the complex ATM standard. The result is that software size increases enormously while forecast units remain modest. In sum, the "sweet spot" for ATM appears to be with the narrow application focus and at the low end, below $25,000/unit.

Viewed in retrospect, the development of the ATM market has been something of a surprise. The ATM standard came out of the telephony side (International Telecommunication Union Telecommunications Standardization Sector, ITU-T, formerly known as the Consultative Committee for International Telegraphy and Telephone, CCITT) of the standards process and the selection of the cell size (53 octets) was seen as a technopolitical concession to that constituency. (Had it been up to the data communications constituency, the cell size would have been larger, perhaps much larger.) Accordingly, in the late 1980s, the expectations were that implementation of this new broadband standard, which finally legitimized packet-switched voice, would be led by the telco giants: AT&T, Siemens, Alcatel, NEC, Fujitsu, Ericsson. In fact, this has not been the case.

With the collapse of the Iron Curtain and the vigorous growth of the deregulatory trend, the telephone equipment side of the house has seen rapid growth in the facsimile and land mobile categories, and incremental improvement via software. The gradual spread of ISDN and Signaling System 7 has opened the door to imaginative and highly useful business services such as 800 and 900 numbers, automatic notification of inward dialed number (ANI), automatic call distribution (ACD) centers, and voice messaging services. Although packetized voice would allow the telcos to combine both logical multiplexing (fast packet switching) and physical multiplexing (combinations of time division and space division multiplexing) to achieve even greater operational efficiencies, given the capital costs of new switches and an underused fiber-optic infrastructure, one can argue that ATM does not make economic sense in the current voice milieu.

Contrast the voice situation with that of data. The ubiquitous Ethernet, which came out of Xerox's Palo Alto Research Center (PARC) in the 1970s, is a 10-Mbps shared medium that on the average delivers about 500-Kbps throughput. Token rings at 4 and 16 Mbps and FDDI I & II at 100 Mbps offer only incremental improvements. None of these technologies, even with considerable compression, will handle high-resolution graphics (800 × 1,200 pixels × 32-bit color × 30 frames/sec), supercomputer channels (100 Mbyte/s and up), or various grades of high-definition digital video (6 to 8 Mbps). Clearly, the popular choices for high-speed data communications, even for the local area, were wanting. Combine that need with a dynamic market dominated by young companies, many less than a decade old, with experience in producing "smart" multiplexers, bridges, routers, and hubs—many with product cycles under two years—and one has today's ATM market for LANs.

A March 1995 product evaluation of ATM LAN switches conducted by Robert Mandeville of the European Network Laboratories, a vendor-independent testing organization in Paris, was symbolic of the dynamics at the leading edge of the ATM market [3]. Fourteen companies were solicited for the product test. Among the negative response were the giants: Fujitsu Network Switching of America, General Datacomm Inc., Hughes Network Systems, NEC America, IBM, Northern Telecom, and Network Equipment Technologies. Accepting the challenge were Bay Networks Inc., Cisco Systems Inc., Digital Equipment Corporation (DEC), FORE Systems Inc., Lightstream Corporation, Newbridge Networks Corporation, and 3Com Corporation. Although quite probably a very imperfect predictor of market share even two or three years hence, it is nonetheless highly indicative of the current furious pace at the low end of the ATM market.

A recent market survey, published in September 1996 *Network World* magazine, identified some 38 switches by 27 vendors [4]:

- ADC Kentrox;
- ADC Telecommunications Network Services;
- Agile Networks, Inc.;
- Alcatel Telecom;
- Ascom Timeplex, Inc.;
- Cabletron Systems, Inc.;
- Cascade Communications Corporation;
- Cisco Systems, Inc.;
- Connectware, Inc.;
- Digital Equipment Corporation;
- First Virtual Corporation;
- FORE Systems, Inc.;
- General DataComm, Inc.;
- Hitachi Telecom USA, Inc.;
- Hughes Network Systems, Inc.;
- IBM;
- Lucent Technologies, Inc.;
- Madge Networks, Inc.;
- NEC America, Inc.
- Newbridge Networ;ks, Inc.
- Northern Telecom, Inc.;
- Scorpio Communications;
- 3Com Corporation;
- Telematics International, Inc.;

- Thomson-CSF Enterprise Networks;
- UB Networks, Inc.;
- Whitetree, Inc.

5.9 ATM SWITCH OVERVIEW

Table 5.2 contains information on these vendors from the same source. The table provides information on ATM switches by application area, configuration/switching architecture, interfaces, throughput, features, interoperability and U.S. list price. As the market is extremely volatile, everything—including the vendor's independent existence—is subject to change.

5.10 SUMMARY

What observations can one draw from this mid-1995 lineup? First, overwhelmingly, the historical orientation. LAN equipment vendors, smart multiplexer makers, packet and frame switch producers are all determined to retain market share via ATM "line extensions." Perhaps less committed but interested nonetheless are the central office and telco equipment manufacturers. Second, the excitement remains with the "startups" (like FORE and Cascade), the byzantine takeovers (Lightstream and StrataCom by Cisco, Centillion by Bay, a merger of Wellfleet and SynOptics, BBN's ATM switch becomes Agile Networks' ATMizer) and collaborative efforts (NT and FORE, IBM and Cascade). Third, it is clearly not in the realm of consumer electronics nor likely will it be for years. It has taken 20 years for Ethernet, invented in the 1970s, to be a million unit per month item for the 3Com Corporation. A similar decent interval is likely before one finds ATM NICs in PCs, personal digital assistants, TVs and portable phones. Last, were it not for the chastening example of Ethernet, where a great many producers found profitable niches, one would be tempted to observe that the ATM switch market is like the auto industry before the Model T Ford—that there are too many producers and not enough volume. It is a reasonable expectation that as the ATM switching market becomes more mature, more sophisticated customers will expect many more software-based features (network and performance management, chargeback, etc.) that are expensive to provide and predicated on a large and growing installed base. At that point, those with a small market share may be vulnerable and need to niche seek or sell out.

Table 5.2(a)
ATM Switches

Vendor	Product	Type					Configuration					Max number of ports per interface type (1)					
		ATM workgroup only	ATM and LAN workgroup	ATM edge switch	Private ATM backbone	Carrier ATM backbone	Fixed configuration	Modular multislot chassis	Modular multitrack system	Hot-swappable ATM modules	Redundancy features	T-1	25M bit/sec	T-3	100M bit/sec TAXI	155M bit/sec OC-3	622M bit/sec OC-12
ADC Kentrox (503) 643-1681	AAC-3 ATM Access			×					×	×	×	28		7		7	
ADC Telecommunications Network Services (800) 336-3891	Cellworx AMS 5001					×		×		×	×	60		30		15	
Agile Networks Inc. (508) 263-3600	ATMizer 125		×		×			×		×	×					5	
Alcatel Telecom (972) 996-5000	1000 AX	×		×	×	×		×		×	×	6,944		1,736		1,736	1,736
Ascom Timeplex Inc. (201) 391-1111	DS-2010				×	×			×	×	×					4	
Cabletron Systems, Inc.	9A000, SFCS 200BX			×	×	×		×	×	×	×			24 or 96	24 or 96	24 or 96	24 or 96
Cascade Communications Corp. (508) 692-2600	500 ATM Switch			×	×	×		×		×	×			112		56	14
Cisco Systems, Inc. (800) 859-2726	Lightstream 1010	×		×	×	×				×	×			16		32	8
	StrataCom IGX			×	×			×		×	×	240		23		6	
	StrataCom BPX			×	×	×		×		×	×	7,680‡		144		96	24

Company	Product														
Connectware, Inc. (214) 907-1093	CELLerity 6120-212 and 6135-312	×		×	×									68	68
	CELLerity 6140-416	×		×	×		68						68		68
CrossComm Corp. (508) 481-4060	XLX ATM Switch	×	×	×	×				32			32		
Digital Equipment Corp. (800) 457-8211	GIGAswitch/ATM System	×		×	×	52			52			52		13
First Virtual Corp. (408) 567-7200	V-Switch	×		×			20					2		
Fore Systems, Inc. (888) 404-0444	ForeRunner ASX-299WG	×		×	×		24			24		16		16
	ForeRunner ASX-200BX and ASX-1000	×	×	×	×	24	24	16	24		16		4
General DataComm, Inc.	GDC Apex ATM switches		×	×	×	30			30	30		30	
Hitachi Telecom USA, Inc.	AMS 5001		×	×	×	60					15	
Hughes Network Systems, Inc. (301) 601-4299	BX5000	×	×	×	×			16		16		16
IBM (800) 426-2255	8260 Switching Hub	×	×	×	×		168	28	56		28
	8285 Nways Switch	×	×	×			48	6	12		7
Lucent Technologies, Inc. (888) 458-2368	GlobeView-2000 Broadband System	×		×	×			384			128

Notes: * When fully configured for specified interface   † Out-of-band only   ‡ Via external concentrators   § In-band only   CMIP: Common Management Information Protocol PNNI: Private Network-to-Network Interface   TAXI: Transport Asynchronous Transmitter/Receiver   TFTP: Trivial File Transfer Protocol   UNI: User-Network Interface *Source:* Mier, E., "It's Arrived," *Network World*, Sept. 16, 1996, pp. 91–100. © 1996 Network World, Inc., Framingham, MA, 01701. Reprinted with permission.

Table 5.2(a) (Continued)

Vendor	Product	Traffic types and signaling							Management	Price
		Constant bit rate (CBR)	Variable bit rate (VBR)	Available bit rate (ABR)	Unspecified bit rate (UBR)	Switched virtual circuit	UNI 3.0 or 3.1	PNNI		
ADC Kentrox (503) 643-1681	AAC-3 ATM Access Concentrator	x	x		x				SNMP, telnet, TFTP	From $20,000 for typical configurations
ADC Telecommunications Network Services (800) 336-3891	Cellworx AMS 5001	x	x		x		x		SNMP (2)	From $33,476; $82,000 for basic system with full management
Agile Networks Inc. (508) 263-3600	ATMizer 125	x		x	x	x	x		SNMP, TFTP	$30,300 for base chassis; $2,500 per OC-3 interface
Alcatel Telecom (972) 996-5000	1000 AX	x	x	x	x	x	x	x	CMIP[+](2)	Vendor will provide pricing for specific configurations only
Ascom Timeplex Inc. (201) 391-1111	DS-2010	x					x		Not specified	Vendor will provide pricing for specific configurations only
Cabletron Systems, Inc. (603) 337-9400	9A000, SFSC 200BX, and SFSC 1000		x		x	x	x	x	BOOTP, SNMP, TFTP	From $17,225–32,250 for base systems
Cascade Communications Corp. (508) 692-2600	500 ATM Switch	x	x	x	x	x		x	SNMP, telnet, TFTP	From $25,000 for base system

Company	Product							Management	Pricing
Cisco Systems, Inc. (800) 859-2726	Lightstream 1010	x	x	x	x	x	x	SNMP, telnet TFTP	From $19,000 for abse chassis; from $2,100 for OC-3 interface
	StrataCom IGX	x		x	x		x	SNMP	From $35,000 for base system
	StrataComBPX	x		x	x		x	SNMP	From $27,500 for base system
Connectware, Inc. (214) 907-1093	CELLerity 6120-212 and 6135-312	x	x	x	x		x	BOOTP, SNMP§, telnet, TFTP	From $9,600 or $11,900 for base system; from $8,500 for OC-3 interface
	CELLerity 6140-416		x	x	x		x	BOOTP, SNMP(4), telnet, TFTP	From $8,800 for base system
CrossCom Corp. (508) 481-4060	XLX ATM Switch	x	x	x	x		x	SNMP	From $19,995 for base system
Digital Equipment Corp. (800) 457-8211			x	x	x		x	BOOTP, SNMP§, telnet, TFTP	$5,250 for 5 slots; $15,240 for 14 slots; from $5,700 for ATM interface
First Virtual Corp. (408) 567-7200	V-Switch	x	x	x	x		x	SNMP§	From $5,700 for base system
Fore Systems, Inc. (888) 404-0444	ForeRunner ASX-299WG	x	x	x	x		x	SNMP, telnet	$9,995 for base system
	ForeRunner ASX-200BX and ASX-1000	x	0+	x	x		x	SNMP	From $15,950 for base system
General DataComm, Inc.	GDC Apex ATM switches	x	x	x	x		x	SNMP, telnet, TFTP	From $6,000 for base system
Hitachi Telecom USA, Inc.	AMS 5001	x	x	x				SNMP‡	From $33,476; $82,000 for base system with full management
Hughes Network Systems, Inc. (301)m 601-4299	BX5000	x		x	x			SNMP	From $26,300 for base system
IBM (800) 426-2255	8260 Switching Hub	x	x	x	x		x	SNMP, telnet, TFTP	From $21,585 for base system
	8285 Nways Switch	x	x	x	x		x	SNMP, telnet, TFTP	From $6,995 for base system
Lucent Technologies, Inc. (888) 458-2368	GlobeView-2000 Broadband System	x	x	x	x			CMIP, SNMP‡, telnet	Vendor will provide pricing for specific configurations only

Table 5.2(b) ATM Switches

Vendor	Product	ATM workgroup only	ATM and LAN workgroup	ATM edge switch	Private ATM backbone	Carrier ATM backbone	Fixed configuration	Modular multislot chassis	Modular multitrack system	Hot-swappable ATM modules	Redundancy features	T-1	25M bit/sec	T-3	100M bit/sec TAXI	155M bit/sec OC-3	622M bit/sec OC-12
Madge Networks, Inc.	Collage 740 Backbone Switch	x			x			x		x	x					16	
	Collage 250 and 280		x				x			x	x					12	
NEC America, Inc. (214) 518-5000	Atomnet/M7			x	x			x		x	x	62		32	62	62	16
Newbridge Networks, Inc. (800) 343-3600	36170 MainStreet				x	x			x	x	x			180		180	
	Vivid ATM Workgroup Switch	x					x									12	
Northern Telecom, Inc. (800) 466-7835	Magellan Passport			x	x			x		x	x	42		18		6	
	Magellan Concorde				x	x		x		x	x			64		64	16
	Magellan Vector			x	x				x	x	x	48		32		32	8
Scorpio Communications	Stinger 1 and Stinger 5			x	x			x		x	x		24 or 96	8 or 32		8 or 32	
3Com Corp. (800) 638-3266	ONcore			x	x			x		x	x		168		56	8	
Telematics International, Inc.	NCX 1E6			x	x			x		x	x	40		16		16	
Thomson-CSF Enterprise Networks (408) 452-0555	ThomFlex 5100			x	x			x		x	x	16	96	16		60	
UB Networks, Inc. (800) 777-4526	GeoSwitch/25	x				x		x					48				
	GeoSwitch/155		x				x									4	
Whitetree, Inc. (415) 855-0855	WS2500 Workgroup Switch	x					x		x							2	

Table 5.2(b) (Continued)

Vendor	Product	Constant bit rate (CBR)	Variable bit rate (VBR)	Available bit rate (ABR)	Unspecified bit rate (UBR)	Switched virtual circuit	UNI 3.0 or 3.1	PNNI	Management	Price
		Traffic types and signaling								
Madge Networks, Inc.	Collage 740 Backbone Switch	x	x	x	x	x	x	x	SNMP, telnet, TFTP	From $13,495/base system
	Collage 250 and 280	x	x	x	x	x	x	x	SNMP§	From $6,995/base system
NEC America, Inc. (214) 518-5000	Atomnet/M7	x	x		x	x	x	x	SNMP, telnet	Vendor to specify pricing
Newbridge Networks, Inc. (800) 343-3600	36170 MainStreet	x	x	x	x	x	x	x	CMIP, SNMP§, telnet	Less than $30,000/ base system
	Vivid ATM Workgroup Switch	x	x	x	x	x	x	x	SNMP, telnet	$23,700
Northern Telecom, Inc. (800) 466-7835	Magellan Passport	x	x		x		x	x	SNMP, telnet	$50,000 typical (voice, video, frame relay and ATM)
	Magellan Concorde	x	x	x	x	x	x	x	SNMP§	Vendor specfies pricing
	Magellan Vector	x	x	x	x	x	x	x	SNMP	Vendor specifies custom pricing
Scorpio Communications	Stinger 1 and Stinger 5	x	x	x	x	x	x	x	BOOTP, SNMP, TFTP	From $15,800/base system
3Com Corp. (800) 638-3266	ONcore	x	x	x	x	x	x	x	SNMP§, telnet	From $26,990/base system
Telematics International, Inc.	NCX 1E6	x	x	x	x	x	x	x	SNMP§, TFTP	From $60,000/base system
Thomson-CSF Enterprise Networks (408) 452-0555	ThomFlex 5100	x	x	x	x	x	x	x	BOOTP, SNMP§, TFTP	From $16,950/base system
UB Networks, Inc. (800) 777-4526	GeoSwitch/25	x	x	x	x	x	x	x	SNMP, telnet	From $6,995/base w 12 interfaces
	GeoSwitch/155	x	x	x	x	x	x	x	SNMP, telnet	From $8,999/base
Whitetree, Inc. (415) 855-0855	WS2500 Workgroup	x	x		x	x	x	x	SNMP§	From $6,995/base system

End Notes

[1] Other compilations of NIC suppliers can be found in Broadband Publishing Corporation, *The ATM Report*, Vol. 3, No. 10, Dec. 30, 1995, p. 5, and in Mier, E., "Buyer's Guide," *Network World*, Sept. 16, 1996, p. 100.

[2] See Bruno, C., "Internet Health Report: Condition Serious," *Network World*, Sept. 23, 1996, pp. 1, 104–111, and Bruno, C., "Fixing the 'Net," *Network World*, Sept. 16, 1996, pp. 57–59.

[3] Mandeville, R., "The ATM Stress Test," *Data Communications*, March 1995, pp. 68–82.

[4] Mier, E., "It's Arrived," *Network World*, Sept. 16, 1996, pp. 91–100.

ATM LAN Switches   6

The ATM LAN market has been characterized as the market "that can't wait." It has been driven by two overlapping sets of customers—applications development groups and laboratories, and leading edge business users with serious LAN performance problems.

In particular, product development and laboratory groups need high bandwidth for data applications immediately. As long as the costs are not too high ($50,000 to $75,000 for an ATM switch and a half-dozen ATM communications adaptors), they can afford to overlook the deficiencies of the current products. Chief among these many short comings are as follows:

- Nonproprietary or proprietary ("prestandard") support of LAN emulation;
- Lack of interoperability with other ATM switches;
- Nonstandard and/or immature congestion control schemes;
- (Early) lack of switched virtual circuits (SVCs);
- Nonsupport of the user-network interface (UNI) 3.0 signaling;
- Nonsupport of the private network-to-network (PNNI) Phase 1 protocols;
- A narrow selection of workstation ATM adaptors;
- Primitive network management.

What did these early LAN switches provide? Mostly speed, specifically rapid transmission and low latency. For those who need to push the edge, the choice is OC-3c at 155 Mbps [1]. Leading edge vendors promise OC-12c at 622 Mbps. To link distributed sites, most vendors offer telco-oriented slower interfaces such as T1 (1.544 Mbps), E1 (2.048 Mbps), E3 (34 Mbps), or T3 (45 Mbps). ATM service options are usually limited to constant bit rate (CBR) and unspecified bit rate (UBR). Most of the boxes handle up to 4, 12, or 16 OC-3c ports. Designed to be purchased for lab use, they do not have accounting capabilities. Control software is strictly first-generation. Further, although the switches are nonblocking they vary widely in their ability to gracefully mix CBR and UBR traffic and maintain performance.

The second, and much larger, group of customers for the early ATM LAN switches are business users, often in the information-beset financial field, frustrated with the limitations of their present LAN technology. They have experienced some or all of the following symptoms:

- Congested Ethernets or token rings;
- Clogged file and application servers;
- Poor or highly variable performance on high-bandwidth applications;
- Increasing difficulty in administering shared media LANs;
- No clear "killer app[lication]" but rather high growth rates in applications, applications bandwidth, and numbers connected to the LANs;
- Growth rates that do not scale well to the present LAN technologies.

Today the cheapest and most standard palliatives involve F/CDDI, 100-Mbps *fast Ethernet,* Ethernet switches, or some combination of the three. Nonetheless, all of these approaches scale poorly and do not meet the bandwidth demands of high-end applications. From a business perspective, they are better rationalized as buying time until the rambunctious LAN-ATM market stabilizes a bit.

In the business arena, the LAN switch market can be characterized as one where the social and physical distances between the user and LAN provider are short and often unsweet. The LAN provider is under heavy pressure to do something positive and to do it fast. The environment is usually customer-owned with a shielded twisted pair/unshielded twisted pair (STP/UTP) infrastructure with multimode fiber (MMF) connecting hubs and routers. Support for the relatively cheap 25-Mbps network interface cards (NICs) come primarily from this constituency. Network management is less important than "plug 'n play" now, and standards less important than functionality now. Too, the client/server-based applications being targeted are often the business' newest and are assigned a high priority. Chargeback schemes are rare and if the payback appears swift, spending under $50,000 is as unbureaucratic as it gets. The attitude is apropos of the Nike sneaker slogan—"Just Do It." The mindset is "fix the problem and spare me the details."

On the credit side, the ATM LAN vendors have been close to their customers and have worked hard to accommodate the installed base (typically, UNIX, TCP/IP, Ethernet, hubs, twisted pair wiring). Most allow a mix and match of Ethernet and ATM workstations and convert the two back and forth, and some support automatic port configuration. ATM port connectors are typically RJ-45. As noted, for the low end, there is weak support for the inexpensive 25.6-Mbps ATM standard for workstations and T1 for WAN connectivity. Reflective of contemporary LAN environments, there is general support of the simple network management protocol (SNMP) and the most popular management information bases (MIBs I & II).

The ATM switches touted for the LAN market offer a subset of the full ATM capabilities. They are high-speed data switches first and foremost and LAN emulation (LANE) is what they do. For lagging LAN performance, they are the promised silver bullets. They offer

- Partial ATM software functionality, effectively CBR and UBR;
- High LAN interface variety and transparency for the Ethernet/TR desktop installed base;
- Server connections at 100 and 155 Mbps;
- LAN interswitch connections at 155 Mbps;
- WAN connections at T1 (1.5 Mbps), E1(2 Mbps), E3 (34 Mbps), T3 (45 Mbps) and OC-3 (155 Mbps).

For the business users, more vulnerable to the economic effects of outages and dependent on vendor support, there are considerable risks with the current ATM LAN market. The state of the standards on ATM are such that proprietary implementations are necessary to do useful work. As all the switch makers are extremely standards-sensitive, both the frequent software releases and hardware swaps to achieve standards compliance and increase interoperability may prove disruptive to applications. For instance, a vendor's proprietary LAN emulation may work better than the initial release of a standards-compliant version. Similar to the PC industry, an ironic counterflow to the uncertainty introduced by the dynamic of standards and new software releases is the certainty of rapidly falling prices and buyers' remorse ("If I had only waited six months!").

The metaphorical sharks are everywhere. In the operations arena, many of the switches produce minimal information about what is going on relative to configuration, network management, or user accounting. The total hardware and software cost of the vendor-provided network management packages is likely to exceed the base cost of the ATM switch. Integration with popular network management packages such as HP OpenView, IBM NetView, and Sun Netmaster is just beginning. Hard-to-spare staff competent with TCP/IP or Netware and shared-media connectionless technology must receive training in ATM fundamentals. ATM support equipment, which encompasses line testers, protocol analyzers,

and ATM simulators, run the gamut from a few thousand dollars to Rolls
Royce cost levels. With the low-end switches packaged with the primary
view to price, hardware fault tolerance may be below expectations. And
the high rate with which the startups are being swallowed by their larger
competitors in the local area—the hub and router vendors—may render
vendor service problematical.

6.1 FOUR APPROACHES TO LEGACY LANs

One can identify four principal ATM LAN internetworking "architectures." The first is "edge routing," enabled by IETF RFC 1483. RFC 1483
specifies the details of protocol encapsulation of LAN packets, both layers
2 and 3. Within the 1483 approach, there are two options. There is the
logical link control/subnetwork access protocol (LLC/SNAP), where the
protocol data unit (PDU) is identified in the header and one can have multiple protocols in the same virtual connection. Another is "VC multiplexing," where each protocol uses a separate virtual connection. The
advantage of 1483 is that one can increase backbone network performance
via ATM. The disadvantages, however, are many. End-to-end performance is still constrained by routers; router management complexity and
cost remain; and 1483 does not accommodate ATM-attached hosts.

The second internetworking approach is the ATM Forum's LAN
emulation or LANE. With LANE, the services of single Ethernets or token
rings are emulated with specialized servers converting LAN MAC addresses to ATM addresses. As LANE emulates but single LAN segments,
so-called one-armed routers (ATM↔MAC) are necessary to logically
bridge multiple LAN segments. But with LANE, certain problems are
solved. LANE accommodates LAN-to-ATM edge devices (bridges) and
ATM-attached hosts; geographic flexibility is achieved. However, problems remain. There is still a router bottleneck between emulated LANs,
bridges are subject to broadcast problems, and the static associations between the host ports and the LAN emulation server limits moves, adds,
and changes.

The third internetworking approach uses IETF's classical IP over
ATM (RFC 1577) to achieve network-level emulation. Logically, the network layer is mapped directly to ATM by, in theory, an address resolution
protocol (ARP) server. ATM-attached hosts are accommodated and they
use 1483's LLC/SNAP encapsulation. The advantages are several. At the
network level, there is more efficient protocol operation, ATM speeds
reach the desktop, and again one obtains geographic flexibility for ATM
workgroups. Of the difficulties, the principal one is that it is limited to
IP, thus excluding the huge Novell IPX constituency. Routers are still

required to link ATM subnets and legacy LANs are connected through routers. Another possible disadvantage is linked to 1577/1483 implementation: some implementations have not used ARP servers but instead PVCs with addresses in NICs or routers, thus requiring a fully connected mesh of PVCs.

The fourth approach may be the "virtual router." It depends on the ATM Forum's multiprotocol over ATM or MPOA (due for a straw vote in December 1996). MOPA will map the LAN's network layer directly to ATM for multiple protocols. In doing so, it will use LANE for layer 2 operations (bridging) and will leverage the IETF's routing over large clouds (ROLC) and next hop routing protocols (NHRP). Logically, the virtual router (with route server) is connected to LAN subnets via SVCs; communications will flow from anybody to anybody without routers. The advantages are signal. The virtual router model supports both ATM-attached devices and legacy LANs. Multiple protocols are supported. Router bottlenecks and administration are removed. The major disadvantage is that although prestandard implementations are here today—such as Newbridge's—MPOA is a new and evolving standard.

6.2 FORE SYSTEMS

FORE Systems, a startup company headquartered in Warrendale, PA, has skillfully ridden the front of the ATM wave. Founded by Francois Bitz, Onat Menzilcioglu, Robert Sansom, and Eric Cooper, all one-time researchers at Carnegie Mellon University in Pittsburgh, Cooper currently serves as president and CEO. Currently with 60% of the ATM switch market, the company remains a David among relative Goliaths [2]. In 1995, it still had less than 1,000 employees and annual sales under $150 million.

FORE initially focused on what might be characterized as the "Lab/LAN" constituency. Their first product, the ASX-100, provided ATM-based data transmission at a variety of speeds greater than the Ethernet, token ring, or FDDI alternatives at costs less than $75,000 for a small lab. The early applications supported were computer data, often involving combinations of video or high-resolution graphics. Ninety-five percent of their current ATM traffic, according to FORE, runs as AAL 5.

FORE's success and small size has led them to an aggressive strategy for both acquisitions and alliances. In the high-capitalization WAN market, they have allied with Northern Telecom (Nortel), essentially hosting their ForeThought software on Nortel's Magellan Passport and Concorde ATM switches. They have a strategic partnership with Cabletron Systems in the LAN switching hub arena. They have moved by acquisition to acquire an Ethernet workgroup switch maker (Applied Network

Technology), a routing and ATM software provider (RainbowBridge Communications), an ATM mux vendor (CellAccess Technology), and an enterprise LAN switching hub maker (Alantec) [3].

Today's FORE's product line includes a number of complementary products for ATM workgroup, LAN backbone, legacy LAN access, and corporate backbone/WAN access, as well as a full line of workstation NICs and an audio/video adaptor. The FORE strategy, including the WAN alliance with Nortel, can be seen in Figure 6.1. In the sections following, FORE's virtual LAN approach, the ASX-200WG, their current workgroup switch, and the role of FORE's ForeThought software will be overviewed.

6.3 THE VIRTUAL LAN APPROACH

The high speeds offered by the ASX-200 enable what are called "virtual LANs," where work groups appear to be in close proximity when in fact they may be physically dispersed over a large building, a campus, or, with the availability of expensive high-speed links, time zones distant. The virtual LAN concept envisions a linked work group (while limiting broadcast messages) along with eased network administration and increased security. Many such virtual LANs are common in today's Ethernet-linked workplaces, primarily to avoid the costs of expensive and time-consuming physical moves. (Unresolved is whether an order of magnitude increase in the bandwidth connecting these workgroups and a second generation of collaborative software and hardware will make the virtual workgroup as effective as "being there.") A notably disinterested portion of the trade press has recently proclaimed that desktop videoconferencing will become the next "killer app[lication]." Some credence for the claims lie in the newly proclaimed desktop videoconferencing standard. There is an ITU-T H.320 standard with H.261 video codec, but the choice of two video formats, common intermediate format (CIF) or quarter common intermediate format (QCIF), and the further choice of three audio compression standards—G.711, G.722 and G.728—renders its popularity problematic. Nonetheless, as several generations of videotelephones have flopped, or achieved very limited acceptance, and competing and noninteroperable schemes are still being heavily marketed, this assertion must be taken with a mountain of salt.

FORE has shipped over 300 units of its initial offering and clearly done a number of things right. It has paid close attention to the installed equipment of the typical lab and focused on the UNIX workstation, Ethernet, TCP/IP environment. Via its ForeThought internetworking software, it delivers four important capabilities: LAN emulation, virtual workgroups, Internet protocol (IP) over ATM, and switched virtual circuits (SVCs). It is one of the few ATM makers to offer network adaptors

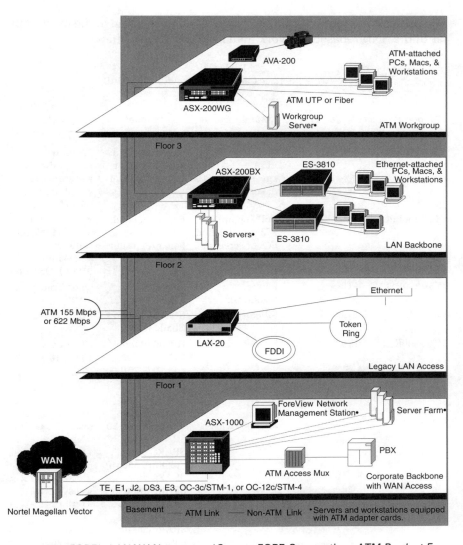

Figure 6.1 FORE's LAN/WAN strategy. (*Source:* FORE Corporation, *ATM Product Family*, Warrendale, PA: FORE Corporation, Oct. 1995, unpaged.)

for a wide variety of popular workstations (see Table 6.2). Aware that few environments can afford a wholesale replacement of their network adaptors, FORE offers a LAN access switch (the LAX-20) to bridge the Ethernet/token ring/FDDI constituency onto the ATM-based LAN. And realizing that most of its target environments use the simple network management protocol (SNMP), it has offered its SNMP-based switch

management system ("ForeView") as a loadable module to the most popular of the SNMP-based network managers, HP's OpenView.

Table 6.1
Network Interface Cards (NICs) Provided by FORE

Adapter Product	Bus Type	Manufacturer	Model	Device Drivers Available*
ESA-200EPC	EISA	Fully compliant EISA specification PCs or servers	Recommended: 486DX/2-66 or better (Pentium class)	Windows NT 3.5 or 3.51; Novell Netware 3.12, 4.x; Windows 95(1.0);OS/2(3.0)
PCA-200EPC	PCI	Fully compliant PCI specification PCs or servers	Recommended: 486 DX/2-66 or better (Pentium class)	Windows NT 3.5, 3.51; Novell Netware 3.12, 4.x; Windows 95(1.0), OS/2(3.0)
PCA-200EMAC	PCI	Power Macintosh	9500, 8500, 7500, 7200	Open Transport 1.1
ESA-200E	EISA	Silicon Graphics	(Power) Indigo 2 Challenge M	IRIX 5.3, 6.2
HPA-200E	EISA	Hewlett Packard	HP9000 Model: 715, 720, 725, 730, 735, 750, 755 J series	HP-UX 9.x, 10.01, 10.10
SBA-200E	SBus	Sun Microsystems	SPARC 1, 1+	SUNOS 4.1.3/Solaris 2.4, 2.5
			SPARC 2, 10, 20	SUNOS 4.1.3/Solaris 2.4, 2.5
			SPARC 5	—/Solaris 2.4, 2.5
			SPARC 630MP, 670MP, 690MP	SUNOS 4.1.3/Solaris 2.4, 2.5
			SPARC 1000	SUNOS 4.1.3/Solaris 2.4, 2.5
			SPARC 2000	SUNOS 4.1.3/Solaris 2.4, 2.5
			SPARCstation IPC, IPX, LX	SUNOS 4.1.3/Solaris 2.4, 2.5
			SPARCclassic	SUNOS 4.1.3/Solaris 2.4, 2.5
			UltraSPARC models	—/Solaris 2.4, 2.5
GIA-200E	GIO	Silicon Graphics	Indigo Indy Challenge S	IRIX 5.3, 6.2
VMA-200E	VME/VME64	Silicon Graphics	(Power) Challenge L, XL, DM (Power) Onyx	IRIX 5.3, 6.2

Adapter Product	Bus Type	Manufacturer	Model	Device Drivers Available*
MCA-200E	MicroChannel	IBM	RS6000 Models: 320, 340, C10, SP2, 3AT/3BT, 360, 370, 380, 390, 58H, 59H, 520, 530, 560, 570, 580, 590, 970, 980, 990, 25O, 25S, 25T, 25W, 34H, 37T, 41T, 41W, 42T, 42W, R10, R20, R24, R30, J30, G30	AIX 3.2.x, 4.1x
PCA-200EUX	PCI	IBM	E20, 43P	AIX 3.2x, 4.1x
ForeRunner LE PC	PCI	Fully compliant PCI specification PCs or servers	Recommended: 486 DX/2-66 or better (Pentium class)	Windows NT 3.5, 3.5.1 Novell NetWare 3.12, 4.x Windows 95 (1.0), OS/2 (3.0)
ForeRunnerLE MAC	PCI	Apple Power Macintosh	9500, 8500, 7500, 7200	Open Transport 1.1 or greater

*Device drivers available for all models listed in each adapter category.
Source: FORE Systems, "ForeRunner ATM Adapter Compatibility Matrix," Aug. 1996.

Most of all, FORE has been nimble. Low-end ATM switches have to be fast and cheap. For better or worse, they are competing against high-end routers. Keeping the price down means putting as much switch functionality as possible in hardware. The problem has been that FORE has been shipping equipment since 1992 while a significant number of technical aspects of the ATM standard remain incomplete. Although the ATM Forum, a group of switch manufacturers, and other interested parties that meet regularly to address and resolve the outstanding issues have made good progress, the ATM standard will probably not be essentially complete until 1997, if then. And if ATM proves popular, akin to X.25, it will be invariably extended. In order to implement functional capabilities that customers want today, such as IP multicasting or network-to-network interface (NNI) signaling, it usually means coding the functions in software today and changing it later if necessary.

6.4 THE ASX-200WG (FOR WORKGROUP) ATM SWITCH

There has been a good deal of hyperbole in the ATM market about second-generation ATM switches. If second-generation means a newer model that implements a number of ATM Forum standards and is capable enough to do useful work, then the answer is "maybe." But if second-generation means a switch where the essential functionality is based on standards—as opposed to proprietary software—and is broadly interoperable with other vendor products, then this "second" generation of ATM switching is more like Release 0.5.

FORE's ASX-200WG (Figure 6.2) implements standards-based approaches in the user-network interface (UNI 3.0), signaling (with Q.2931 and the NNI), and in handling LAN traffic (LANE and RFCs 1483 and 1577). Its SNMP-based (v1, v2) network management approach, ILMI, with MIB II compliance and FORE custom MIBs, is arguably about as standard as it gets in that area. Its copper-based media support is standard (CAT-5 UTP and Type 1 STP for 10–155 Mbps) per EIA/TIA 568A, ISO 11801 and ATM Forum's physical medium dependent (PMD) standard (AF-PHY-0015.0000). Standard, too, are the UTP/STP connectors: RF-45, DB-9 and the "Boy George" 4-position data connectors. DB-15 (female) is used to connect Ethernets and multimode fiber employs SC connectors.

Figure 6.2 Front panel view of the ForeRunner ASX-200WG. (*Source:* FORE Systems, "ForeRunner ASX-200WG ATM LAN Workgroup Switch," 1995, unpaged.)

But other critical areas are unabashedly proprietary. FORE's traffic features include source shaping, dual "leaky bucket" policing, forward error control notification (FECN)-based congestion management and multiple priority levels. Although all are from the ATM-approved toolbox for traffic management, there is no guarantee that they will work effectively with other makers' traffic management schemes. Such distinctions between "interoperability"—versus the ability to create multivendor networks that work reliably and efficiently—are the provinces of the ATM cognoscenti.

Whether Release 0.5 or "second generation," the main reason to be excited by the ASX-200WG is that it has enough features to perform useful work in a variety of environments. It offers SVCs as well as PVCs. There are lots of LAN interfaces (100-Mbps transparent asynchronous transmitter/receiver interface [TAXI], 155-Mbps OC-3c and 622-Mbps OC-12c SONET/SDH) and WAN interfaces (1.5-Mbps T1, 2-Mbps E1, 6-Mbps J2, 34-Mbps E-3, 45-Mbps DS-3, 155 OC-3c, and 622-Mbps OC-12c SONET/SDH). The ASX-200WG scales from 12 to 24 ports and connects up to 24 clients. It, and its brother switch, the fault-tolerant ASX-200BX, offer a number of nice to have features for fault tolerance (dual and hot swappable power supplies, hot swappable network modules, redundant fans, environmental monitoring). Too, FORE offers lots of ATM adaptors for workstations: Sbus, EISA, HP EISA, PC EISA, GIO, PCI, VME, Micro-Channel, and Turbochannel. Table 6.2 below provides ASX-200WG technical specifications in greater detail.

Table 6.2
ForeRunner ASX-200WG Technical Specifications

Technical Specifications

System Hardware:

Switching fabric: 2.5 Gbps, nonblocking

Number of ports: 12 to 24
Switch transit delay: 10 microseconds

Control processor: i960CA
Power (nominal): 120/240 VAC @
  60/50/Hz, 3.0/1.5 amps

Dimensions: H: 4.75 in. W: 17.50 in.
D: 18.00 in. H: 12.1 cm W: 44.5 cm D: 45.7
Weight 27.6 lbs (12.5 kg)

Output buffers: maximum of 13,312 cells
  per port

Table 6.2 (Continued)

Technical Specifications

Traffic policing: UPC, dual leaky bucket
  support

Bandwidth management: *Smart Buffers*

Per-VC queuing (maximum 13,312
  queues/ network module)

Early packet discard (EPD), partial packet
  discard (PPD) EFCI flow control

Connection setup time: 10 milliseconds, 100 calls/second
Maximum port speed: 622 Mbps
  (OC-12c/STM-4c)
Ethernet interface: 802.3-compatible,
  DB-15 female connector
Serial interface: DB-9 connector

Front panel indicators: carrier detect, Tx,
  Rx, diagnostic indicators

System Software

ForeThought Internetworking Software:
  Switched virtual circuits (SVCs)

Permanent virtual circuits (PVCs)

Smart permanent virtual circuits (SPVCs)

Multicast and broadcast

Three levels of cell priority with 127 sublevels

ATM Forum UNI v3.0 and SPANS SVC protocols

IISP and SPANS NNI for interswitch connections

ForeView Network Management Software:

Simple Network Management Protocol (SNMP v1)

Point-and-click operation

Automatic discovery and mapping of Forerunner ATM networks

Integration into existing netwrok management platforms

HP OPenView, SunNet Manager, IBM NetView/6000

General Specifications

Standards compliance: ITU I.361 ATM
  Layer; UNI v3.0
Emissions: FCC Part 68, Subpart 15
  Class A; CISPR 22, Class A;
EMC IEC 801-3, Level 2 and IEC 801-4,
  Level 2; VCCI, Class 1
Safety: UL1950; CSA 22.2; IEC950; EN50082-11; VDE Safety

Technical Specifications

Environmental:
Operating temp: 5 to 40C up to 10,000 ft.
Operating humidity: 10 to 90% RH noncondensing
Storage temp: −40 to +70C up to 30,000 ft.
Storage humidity: 5 to 95% RH noncondensing
ESD susceptibility: IEC 801-2

Source: FORE Systems, "ForeRunner ASX-200WG ATM LAN Workgroup Switch," Aug. 1995.

Probably where "FORE the useful," as opposed to "FORE the particularly standard," shines brightest is in the switch and network management area. FORE's ForeView network management software integrates into the big three of the UNIX-based, point-and-click SNMP managers: HP Openview, SunNet Manager, and IBM's NetView/6000. It supports automatic network discovery and mapping. ForeView can create and reconfigure virtual workgroups, monitor network links and devices, and perform inventory management. In sum, it goes a long way toward aiding the configuration and troubleshooting of what is a complex technology. Table 6.3 provides a more complete description of the technical specifications for the ForeView network manager.

In retrospect, there are several interesting observations on the FORE product line vis-à-vis the ASX-200WG. Unlike others, FORE has bundled legacy LAN support into a separate box, their ForeRunner LAX-20 LAN Access Switch, rather than in the 200WG itself. And in a even more controversial move, FORE decided to enter the very competitive Ethernet switching market with their ForeRunner ES-3810 ATM-Ready Ethernet Workgroup Switch, a move that will put them head to head with 3Com and Bay Networks.

6.5 FORETHOUGHT

FORE's current supremacy in the ATM switch market is directly due to its ForeThought operating software. ForeThought fills in the chinks in the current standard, adds attractive features, and (noted above) its ForeView network manager provides enough capabilities to make ATM switching a real thing.

Table 6.3
ForeView Network Management Technical Specifications

Technical Specifications

Hardware:
  SPARCstation or HP700
  32 MB memory
  20 MB free disk space
  64 MB swap space
  Color monitor
  ATM SBA-100/200, HPA-200, or Ethernet
  network interface

Software:

SunOS 4.1.3 or HP-UX 9.x
HP OpenView SNMP Platform version 3.x

Supported ATM Products:
ForeRunner ASX-100 and ASX-200 running
  ForeThought software version 2.2 or later
ForeRunner LAX-20 LAN access switch
  running *ForeThought* software version
  1.2.1 or later
ForeRunner ATM adaptor card family
  running *ForeThought* software version
  2.2 or later

MIB Specifications
SNMP v1 and 2
MIB II compliant
Custom FORE MIBs for:
  Adaptors
  Switch Software, Ports, and Modules
  SONET
  DS-3
  E-3

Monitored Parameters:
  Switch-Based
  Channel (VC) received & rejected cells
  Path (VP) received & rejected cells
  Port (link): received & transmitted cells,
  overflows, errors

Host-Based
Host transmit (TX) and received (RX) cells
AAL0: cell TX, RX, and discards
AAL3/4: cell TX, RX, and discards, PDU
  TX and RX, and discards, AAL3/4 Errors:
  CRC, CS, and SAR protocol
AAL5: Cell TX and RX, and discards,
  PDU TX and RX, and discards, AAL5
  Errors: CRC and CS protocol SONET:
  section BIPS, line BIPS and FEBEs, path
  BIPS and FEBEs, SONET errors: ATM
  correctable and uncorrectable HCS

Configurable Parameters
DS-3 line type
DS-3, E-3, and OC-3 TX clock and
TX/RX scrambling
DS-3
E-3 and OC-3 framing

Source: FORE Systems, "ForeView Network Management for ForeRunner ATM Networks," 1994.

FORE's software organization is arranged in a non-OSI, four-layer model depicted in Figure 6.3.

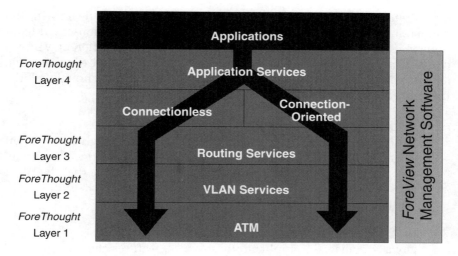

Figure 6.3 ForeThought's layered architecture. (*Source:* FORE Systems, *ForeThought and the ATM Internetwork*, version 2.0 Warrendale, PA: FORE Systems, 1994, p. 5.)

From the bottom up, layer 1 provides ATM transport services. This is where non-ATM traffic is converted into ATM cells, SVCs are set up, and routing occurs. The layer 1 service list includes the following:

- Automatic network configuration;
- P-NNI;
- Bandwidth management;
- UNI 3.x SVCs;
- Load balancing;
- Automatic rerouting.

Layer 2 consists of VLAN services. VLANs, as noted earlier, are logical groupings of users that share a common broadcast domain. Fore-Thought VLANs use LANE and classical IP to support logical domains that spans both ATM-emulated LANs and conventional LANs. The list of layer 2 services include the following:

- Application and protocol transparency;
- ATM Forum LAN emulation;
- Classical IP/MPOA (RFC 1577);
- IP multicast.

Distributed routing services form layer 3. This layer deals with the connectionless legacy, the need to route within a connection-based technology. These cases include communications between different VLANs, conversion between different MAC types such as Ethernet to token ring, and internetworking with existing networks. Hence ForeThought's "distributed" routing services form layer 3:

- Distributed route calculation;
- "Cut-through" routing (which FORE defines as providing the client with a final destination address so that an SVC can cross the subnet boundary directly to the destination host; the effect is that the ATM network, regardless of size, appears as a single network hop);
- Multicast;
- Interoperability with existing routers.

Layer 4 consists of applications services, features that are required on a per application, per session basis. These application services include the following:

- Access control security;
- QoS guarantees (essentially, bandwidth reservation);
- Connection auditing (call records);
- Resource management (VC management).

Using this four-layer schema, of which the middle two layers are consumed with supporting connectionless legacy systems, FORE provides some very revealing layer-by-layer charts (see Tables 6.4–6.7). On the face, they juxtapose FORE's compliance with industry standards and FORE's "added value." There exists an apparent paradox here: FORE is committed to standards compliance *and* feature leadership, with the latter almost always involving proprietary software. Another critical view might be they illustrate the current distance between needed functions that have been standardized and needed functions that have to be met today with proprietary software. They appear, for each of the four layers designated by FORE, like pencil marks and dates on a doorsill—they chart the relative maturity of the ATM standard. And if FORE continues to be nimble, regardless of the state of the standards (the left columns), there will always be customer-enticing, proprietary "added value" (the right columns).

Today's ATM workgroup switches, of which FORE's ASX-200WG is the premier example, can in some ways be compared to a technology that ATM switches are likely to replace—time division multiplexers (TDMs). TDM muxes such as Newbridge and Network Equipment Technologies

Table 6.4
ForeThought Layer 1 Standards Map

FORE's Compliance With Industry Standards	*FORE's Added Value*
Layer 1 ATM Transport Services	
UNI 3.X and Beyond	
ILMI address administration	SPANS UNI
SVC signaling (Q.93B+)	SPANS addressing
Traffic policing (dual leaky buckets)	Anycast
ATM Routing Standards	
I-ISP	SPANS P-NNI
P-NNI	Distributed ATM connection routing
	ForeThought connection admission control
	Auto reroute
	Load balancing
	Smart PVCs
ATM Adaptation Layer Standards (Segmentation and Reassembly)	
AAL5	
AAL3/4	
AAL 1	
ATM layer cell switching	
Cell switching (VPI/VCI)	
Traffic policing (dual leaky buckets)	*ForeThought* bandwidth management
ATM MIB	FORE enterprise MIBs
Quality of service classes (CBR, RT-VBR, NRT-VBR, UBR, ABR)	Per-VC queuing
	Packet-level discard
Physical layer standards	
SONET/SDH, TAXI, Desktop 25, DS3/E3, T1/E1, J2, F-UNI/DS0	
Physical layer MIBs	FORE enterprise MIBs

Source: FORE Corporation, "ForeThought and the ATM Internetwork," version 2, White Paper, 1995, p. 9.

can interoperate in the sense of linking T1s and administering DS-0s. But few would voluntarily intermix different brands of TDM muxes because many of the proprietary features that make them attractive would be

Table 6.5
ForeThought Layer 2 Standards Map

FORE's Compliance With Industry Standards	*FORE's Added Value*
Layer 2: VLANs Services	
LAN emulation (LECS, LES, BUS)	Distributed/redundant LECS, LES, BUS
Classical IP (RFC 1577)	
	Intelligent BUS
802.2 LLC LSAP addressing	Broadcast/multicast reduction (spoofing/caching)
802.1d spanning tree	VLAN by port, MAC L2 address, L3 address, L2 protocol, L3 protocol, host name, or application
802.1d SRT bridging	VLAN validation (authentication)
802.3/802.5 MAC	VLAN roaming

Source: FORE Corporation, "ForeThought and the ATM Internetwork," version 2, White Paper, 1995, p. 13.

Table 6.6
ForeThought Layer 3 Standards Map

FORE's Compliance With Industry Standards	*FORE's Added Value*
Layer 3: Distributed Routing Services	
Route calculation	
Routing protocols (RIP, OSPF, BGP-4 IPX-RIP, NLSP)	Distributed, redundant route calculation (MPOA) servers
MPOA	Integrated ATM internetworking
Protocols and addressing (IP, IPX)	Integrated, scalable multicast
	Integrated IP-RSVP/ATM QoS
IP-RSVP	Integrated NSAP-based addressing
Packet switching	Redundant DHCP servers
DHCP	

Source: FORE Corporation, "ForeThought and the ATM Internetwork," version 2, White Paper, 1995, p. 18.

Table 6.7
ForeThought Layer 4 Standards Map

FORE's Compliance With Industry Standards	FORE's Added Value
Layer 4: Application Services	
ATM API for new applications	Preapplication QoS for existing applications with no program changes
WinSock 2.0	Preapplication, per-user, per-connection security with native ATM performance
	Session records for call detail records (CDR) billing, performance modeling, and security audits

Source: FORE Corporation, "ForeThought and the ATM Internetwork," version 2, White Paper, 1995, p. 21.

compromised, mostly notably in network management. This condition of interoperability/noninteroperability has been taken as a given in the mature TDM market. By contrast, ATM—which can be seen simply, but inaccurately, the next generation smart multiplexer—is striving, through the ATM Forum, for a much broader span of interoperability with a technology that represents an order of magnitude greater complexity. Apropos of whether the ASX-200WG is "second generation" or not is how one marks the scorecard, with the ironical observation that vis-à-vis the progressive elaboration of the ATM standard, we have no idea of how many generations are forthcoming.

6.6 CISCO CORPORATION

Cisco's entry and presence in the ATM switch market is rich with paradox. A phenomena of the 1980s, this multibillion dollar San Jose, California company owns the lion's share of a router market that has stitched together thousands of LAN swatches into world-spanning fabrics. Their strength has been in the permutations of connectionless protocols and in providing multiprotocol routing. Protocol omnivores, they have successfully interconnected with connection-oriented earlier protocols like X.25, and, to IBM's discomfort, SNA. Running with the virtual LAN concept, they have expanded from routers into LAN switches, introducing the high-end Catalyst LAN switch (from Crescendo Communications) in 1994 and adding a desktop/workgroup LAN switch (from Kalpana) later in the same year. With ISDN "B" channel service, they have extended the virtual LAN even to the small office/home office (SOHO) market. From a smart, aggressive company whose self-depreciating humor includes such slogans as "Headers R Us," the internetworked world was their oyster.

So, what does router king Cisco do with ATM, which many view as the successor technology to connectionless virtual LANs? How could they push forward with ATM without cannibalizing their router base? The official response came in February 1994, with "CiscoFusion," the (promised) seamless integration of routing and switching. More tactically, it involved a number of calculated business decisions. The first was securing their rear—the enormous, worldwide installed base of "legacy" Ethernets and router networks. They correctly forecast that this market, based on $99 Ethernet cards and cheap PCs, would continue to grow from the bottom up and many would eventually internetwork. (Cisco was understandably active in the IETF's upgrading of IP to IPv.6, which provides a huge increase in address space while improving security and easing network administration [4].)

For their mid-level, mostly business customers, they met the bandwidth/throughput problems at the workgroup level. Retreating from the Ethernet, shared-media heritage and progressive microsegmentation to maintain performance, they jumped into the switched Ethernet market while their routers supported faster alternatives such as CDDI/FDDI and 100-based T, and, on the WAN side, bandwidths to 155 Mbps [5].

For the high end, they embraced the enemy, buying into both the workgroup (a joint venture with NEC) and edge ATM markets (Lightstream, a BBN-Tandem joint venture). The initial workgroup switch, a partnership of NEC hardware and Cisco software, was the Hyperswitch Model A100, which was renamed the Lightstream 100. (In Spring, 1996, it was succeeded by the more powerful Lightstream 1010.) The BBN-Tandem switch became today's Lightstream 2020. Other ATM components were ATM interface processors (AIP) for their top-of-the line 7000 routers and a comarketing agreement with Litton-Fibercom for the important (and currently expensive) CSU/service multiplexing functions.

Recently, in 1996, seeing the frame relay market taking off, Cisco bought StrataCom, one of the market leaders. They offer the StrataCom IGX, a hybrid frame relay/ATM switch as another edge option for WANs.

In CiscoFusion, Cisco sees low-cost LAN switches combining with the "switched internet" at the WAN. Banking upon the size of their installed base and their formidable protocol expertise, Cisco sees itself as best able to meld these very dissimilar technologies into distance-defying, virtual LANs. More modestly, Cisco can offer Etherswitches for collisionless LAN transmission, routers for protocol segmentation, and frame relay and ATM switches for high-speed WANs.

6.7 CISCO'S LIGHTSTREAM 1010 WORKGROUP SWITCH

The Lightstream was designed from the bottom up for uncertainty. Given the constantly advancing ATM standards, everything in the box (the chassis is the same as that of Cisco's Catalyst TM 5000 LAN switch) is designed-in to be swapped out. This phenomenon—the Sisyphian struggle to keep up with a moving standards target yet use silicon effectively—has been lampooned by industry watchers with the rubric: "ATM: Where the rubber meets the sky!" In the case of the 1010's predecessor, Cisco had to offer 50% turn-in allowances. Compared to their router lineup, where they have repeatedly upgraded their Internetworking Operating System (IOS) to provide additional functions and protocol support, the ATM upgrade path is significantly more complex as high-performance demands hardware swaps (read downtime) as well as new software loads.

Up to now, most of the challenges in distributed computing and networking have focused on software upgrades and were typically performed remotely by the central site. Having branch office personnel doing board swaps, although not unheard of, adds excitement. Realizing that most of the accounts into which the Lightsteam 1010 will be placed already are Cisco router customers, Cisco has stressed the level of product software continuity (officially, CiscoFusion) with the router line. For example, they even offer as an option the same command line interface their routers use. Thus there is continuity, for better or worse, with IOS, despite the credo that "Switching people are layer two-oriented; routing people are layer three-oriented.") In the spring of 1996, the Lightstream 1010 started at about $19,000, with much of the cost in the interface cards. As box sales and the low-end competition heats up, these prices will surely fall. Looking further to the future, Cisco's huge connectionless base forms their soldiers, and multiple protocols over ATM (MPOA)—the future standard promising seamless multiprotocol connectivity with ATM—their battlefield.

6.8 CISCO HARDWARE

The 1010's hardware design is straightforward and, as noted above, is oriented toward easy replacement. There is a five-slot chassis with the option of dual, fault-tolerant, load-sharing power supplies. There are basically two kinds of cards, processor and I/O. Within I/O, there are a variety of interface cards (see Table 6.1). The processor card, called the ATM switch processor module (ASP), comes with lots of fast shared memory for buffers, a nonblocking switch fabric, a 100-MHz MIPS R4600 RISC processor for central intelligence, and a field-upgradeable feature card.

The remaining slots support up to four I/O cards, called carrier modules (CAMs), each of which can support up to two interface cards, called port adaptor modules (PAMs). All of the cards except the processor card are hot-swappable. Cisco has announced that successor models of the 1010 will be stackable.

The port density of the 1010 allows for the following maximum configurations:

- Up to 192 25 Mbps ATM ports;
- Up to 32 SONET ST3c/STM1 155-Mbps ports;
- Up to 8 SONET STS12c/STM4c 622-Mbps ports;

Figure 6.4 summarizes the 1010's physical layer, data rate, media, and connector options.

Cisco claims that a Lightstream 1010, with the maximum memory configuration of 64 MB of DRAM, will support up to 32,000 point-to-point connections and 1,985 point-to-multipoint connections.

6.9 CISCO SOFTWARE

As with FORE's ASX-200WG, Cisco has made every attempt to be as contemporaneous as possible with the latest ATM standards. The Lightstream 1010 supports the ATM Forum's ABR congestion control and the UNI 3.0 signaling protocol (3.1 and 4.0 are futures). Cleverly, all the mechanisms supporting Cisco's QoS (ControlStream), the ATM traffic classes and AAL types, congestion avoidance (VirtualStream), as well as routines for traffic policing, multiple levels of priority, intelligent early packet discard, connection admission control and traffic pacing, are placed on the field-replaceable feature card, waiting for the next standards shoe to drop.

The interim interswitch signaling protocol (IISP), which allows support of both SVCs and PVCs, is part of the "default software image" on the Lightsteam 1010. One rationale for IISP is that it allows backward compatibility with switches that have not yet implemented the private network-network interface (PNNI). Cisco provides what they call a "soft" (or automated) PVC support, which facilities PVC/permanent virtual path (PVP) setup and rerouting by using signaling protocols across the network. Also featured is virtual path (VP) "tunneling" to enable signaling across public networks.

Optionally, in addition to the default IISP, Cisco offers the ATM Forum's PNNI Phase 1. The PNNI provides for the complex address aggregation and routing hierarchy mechanisms that will allow large private ATM

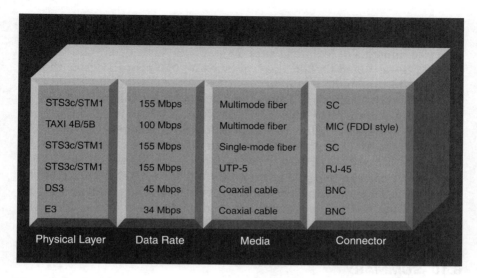

Physical Layer	Data Rate	Media	Connector
STS3c/STM1	155 Mbps	Multimode fiber	SC
TAXI 4B/5B	100 Mbps	Multimode fiber	MIC (FDDI style)
STS3c/STM1	155 Mbps	Single-mode fiber	SC
STS3c/STM1	155 Mbps	UTP-5	RJ-45
DS3	45 Mbps	Coaxial cable	BNC
E3	34 Mbps	Coaxial cable	BNC

Figure 6.4 Physical layer, data rate, media, and connectors. (*Source:* "Cisco Hyper-Switch ATM Family," Cisco Systems, 1994.)

networks. (Cisco's first release is limited to a single peer group.) The PNNI also allows important features like QoS on demand, load balancing, redundant links, and access lists for security. Perhaps even more important, PNNI enables Cisco to offer a "plug 'n play," autonomous operation option in which the PNNI protocols discover switch locations and the ILMI protocols configure the addresses of ATM end systems. Provided, too, is automatic recognition of interface type—UNI or NNI, public or private. Cisco claims the plug 'n play options allow a network of Lightstream 1010s to be deployed with no operator intervention at all. Given the complexity of ATM, Cisco's clever use-out-of-the-box capability is a harbinger of other ATM "autoconfigurations" to come.

The Lightstream 1010 supports a number of features for switch management. ATM F4 and F5 capabilities for operations, administration, and maintenance (OA&M) cells flows are supported. The BOOTP for IP address autoconfiguration, Telnet, and trivial file transfer protocols (TFTPs) are supplied for remote configuration and software download. SNMP monitoring and configuration are handled through two products, the graphical user interface (GUI)-based CiscoView, which provides the chassis physical view, configuration, performance monitoring, and minor troubleshooting. Their AtmDirector product allows for system management of homogeneous (all-Cisco) ATM internetworks. These products, in turn, can run under Cisco's StreamView graphical network management tool, which, in turn, runs under HP's popular OpenView network management platform. All standard ATM MIBs are supported along with various Cisco extensions. In an attempt to leverage their router base—and the large numbers of technical support staff who have been trained in the Cisco IOS—as noted earlier, the Lightstream 1010 also allows configuration via IOS' command line interface (CLI).

For fault diagnosis, Cisco offers several mechanisms to facilitate monitoring via an ATM analyzer. What they call "port snooping" and "connection steering" allow, in concert, for the connections in any given direction of a selected port to be transparently mirrored across a particular monitoring port. An Ethernet port and two serial ports (one that can be used for a local terminal and a second for remote access) are available for monitoring. These monitoring ports are available with several access protections, from multiple password levels to the terminal access controller access system (TACACS) for remote access validation [6].

6.10 SUMMARY

To date, LAN switches have been where the ATM "box sales" are [7]. One industry survey has found that the typical ATM environment consists of two switches and twelve workstations. This toe-in-the-water approach is likely to continue, given the constantly evolving ATM standard. For the LAN switch makers—of which FORE and Cisco are but two of many—the challenge is be able to economically modify their hardware and software to provide the essential customer functions (SVCs, UNI, PNNI, LANE) as well as support the popular interfaces and link rates, keep up with a constantly changing standard, and make money while doing it. Particularly in the LAN market, they are conscious of the tendency of small manufacturer-homogeneous physical LANs to grow into very large manufacturer-heterogeneous virtual LANs. As a result, there is arguably a greater emphasis on standards adherence at the LAN switch level than among edge switches which, as Cisco tactfully observed, "can take advantage of their prestandard capabilities [8]."

End Notes

[1] The 140-Mbps TAXI interface was supported by FORE on their early ASX-100 model workgroup switch; when it was not supported by the ATM Forum it was not included as an option on subsequent models.

[2] Jeffries Research, 1995; quoted in *Network World*, Jan. 30, 1995, p. 16.

[3] *Network World*, Dec. 18, 1995, pp. 10, 14.

[4] An interesting development in the IETF camp involves the Resource Reservation Protocol (RSVP), a layer 3 protocol with the objective of binding QoS across router net-

works. If practical, this would be a further diminution of the distinctions between (connectionless) router- and (connection-oriented) switch-based networks.

[5] "Microsegmentation" refers to maintaining Ethernet performance in the face of larger (or more frequent) broadcasts by putting power users on smaller and smaller segments, ultimately one user per segment à la Etherswitches.

[6] TACACS is an authentication protocol that runs on a security server, generating unique numerical passwords every few seconds. Remote users have a credit card token, or software on their computers, that simultaneously generate the same numbers, enabling access.

[7] This must be contrasted with the other major ATM market segment, the "carrier/WAN market," where the number of units are far fewer but the prices considerably higher (see Chapter 7, ATM Access or "Edge" Switches).

[8] Cisco Product Announcement, I.D. No. 119PA, p. 3.

ATM Access or "Edge" Switches

Most ATM "edge" switches today are data-oriented switches positioned between high-capacity routers and high-capacity bandwidth. With prominent router vendors like Cisco and Bay Networks now in the ATM market, hybrid ATM edge switch/routers are likely. With well-known "smart-mux" vendors like Network Equipment Technologies and Newbridge, and frame relay vendors like StrataCom (now part of Cisco) and Cascade (recently acquired by Ascend), other kinds of intriguing ATM edge products have appeared.

Nor do the combinations stop there. Increasing quantities of voice and video will flow through edge switches. And there are other capabilities coming in the near future. Today's ATM is oriented toward leased or owned bandwidth. ATM cells in WANs flow over session connections called switched virtual circuits (SVCs) that employ, typically, telco-provided DS-1/3s. At some point the SVCs themselves may employ dial facilities, holding out the promise of pay-as-you-go broadband and a much more versatile, and probably cheaper, service. When this happens, edge switches will assume this role as well.

Technically, what distinguishes the ATM edge switch from the LAN ATM switch?

It is more a matter of degree. Edge switches, because they must be able to handle retransmissions in a high-bandwidth environment with significant delays, must have much larger buffering capability than their smaller LAN cousins. These big buffers entail significant amounts of random access memory (RAM) and add to per-port and switch costs. Where LAN switches are trying to deliver broadband to the desktop, edge switches are trying to move aggregated desktop traffic (in broadband quantities) to the wide area within quality of service (QoS) guarantees. Edge switches typically connect with Ethernet hubs or routers on the local area side, and telco resources on the wide area side. Indeed, they are likely to be physically adjacent to the demark as they terminate customer-owned multimode fiber (MMF) and telco copper or single-mode fiber (SMF). Too, when compared to LAN ATM switches, they are likely to offer less variety in LAN interfaces, more variety in WAN interfaces, and greater port densities (more device connectors per interface card). As they manage expensive broadband resources, edge switches are much more likely to be a shared resource. Accordingly, fault tolerance and network management are more important and serious accounting and network management capabilities are expected. (Nonetheless, in these areas most of the edge switches today share the technical shortcomings of their LAN brethren.) All tolled, and not surprisingly, edge switches usually represent a greater investment ($50,000–$150,000).

Economically, the edge switches fit in neatly in the tradition of "cream skimming" the most lucrative markets. Just as in the early days of deregulation MCI and Sprint underpriced AT&T on high-traffic, interurban routes, a number of ex-community access television (CATV) providers and bypass outfits (Metropolitan Cable, Teleport, MFS Datanet [now MFS Worldcom], MCI Metro [now British Telecom]) stand poised to underprice the ex-Bell Operating Companies (BOCs) in the lucrative, high-volume urban markets. MFS Datanet, the first to announce an ATM data service connecting a number of major cities, prices itself roughly 25% below what a comparable, leased DS-3 circuit would cost while offering ATM QoS guarantees. Obviously, large financial concerns that are already employing expensive DS-1s and DS-3s for data transmission can see a similar opportunity with their private networks. Similar in objective to the routers, packet and frame switches, and smart muxes they will replace or subsume, ATM edge switches allow one to wring more throughput from expensive bandwidth—with QoS—by the logical multiplexing of multiple data streams.

7.1 CASCADE COMMUNICATIONS CORPORATION

The Cascade Communications Corporation is a Westford, Massachusetts, startup that, with StrataCom, pioneered the multiservice WAN switch. Incorporated in 1990, it received venture capital in 1991, produced its first product (the B-STDX) in 1993, and went public in 1994 (NASDAQ: CSCC). Today, it offers two complementary hardware platforms, its B-STDX 9000 edge switch and its newer ATM CBX 500 enterprise switch (available in two models). The CBX 500 is complemented by ATM access equipment from Sahara Networks, a start-up vendor bought by Cascade in January 1997. Thus far it has made its mark primarily in the frame relay market selling to carriers and service providers. Its versatile B-STDX offers, in addition to frame relay (FR), integrated service digital network (ISDN) primary rate (PR), switched multimegabit data service (SMDS), and ATM capabilities. Much of Cascade's success has been at the expense of the TDM vendors. With ATM emerging as a credible contender to the FR market, the multiservice approach embodied in the B-STDX represents a calculated bet on the velocity of the evolution to ATM—that it will be incremental, perhaps an application at a time.

Furthermore, Cascade's evolutionary phases (Figures 7.1–7.3) raise intriguing issues about the nature of the "edge" itself. Most network pundits assume that ATM, because of its speed, QoS, and multimedia abilities, will be the technology of choice for the WAN backbone. Most orthodox ATM-dominated views see an ATM edge switch for data, with the addition of a service multiplexer if voice and video are supported. But less clear is how the edge—or edges—will actually evolve, with hybrid routers, hybrid frame switches, and other possibilities. Even in tidy locations where TDM equipment serves as a service multiplexer today, it is likely that all sorts of span-of-control, fault-tolerance, and legacy influences are likely to make the network edge a lot less tidy than the ATM vendors picture it. Cascade may have the most realistic vision as to how rapidly the transition to ATM will actually occur; tellingly, most of its customers seem to be carriers/providers.

7.2 CASCADE'S B-STDX 9000 SWITCH FAMILY

Cascade calls its B-STDX 9000 a "multiservice WAN platform" supporting FR, SMDS, and ATM. As noted above, it represents a hedged view of the role of the WAN edge or edges. (The CBX 500 model is the company's pure-ATM alternative.) With its B-STDX 9000, one gets a sort of "appropriate technology" approach: a large number of disparate legacy systems flow across an ATM WAN core, with applications transitioning to end-to-end ATM only when the requirements demand and the economics permit.

The B-STDX comes with a 16-slot chassis. The basic architecture uses symmetrical multiprocessing and consists of a control processor (CP) module that handles network and system management interacting with multiple input/out (I/O) modules that perform the real-time switching functions. All modules employ the Intel I960 RISC chip.

As appropriate to an edge switch connected to high-cost WAN bandwidth, there are a number of high-availability options. Cascade, like other vendors, has attempted to maximize port densities, with the advantages of slot and cost parsimony but with the disadvantages of multiplying the effects of a single component failure. Accordingly—and perversely, from

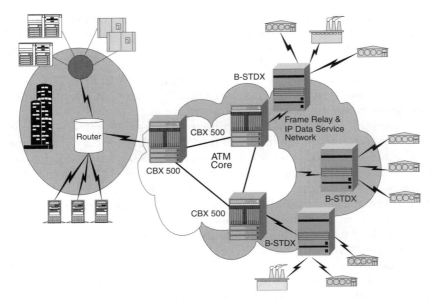

Figure 7.1 Cascade's ATM integration phase 1. (*Source:* Cascade Communications Corporation, "Cascade 500 High Scalability ATM Switch Applications & Technology," Network White Paper, May 1995, p. 4.)

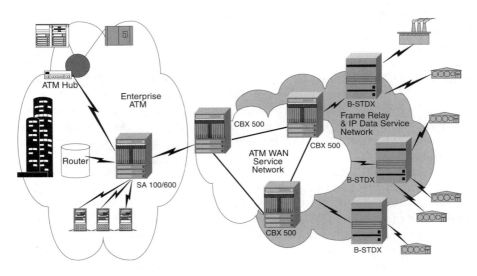

Figure 7.2 Cascade's ATM integration phase 2. (*Source:* Cascade Communications Corporation, "Cascade 500 High Scalability ATM Switch Applications & Technology," Network White Paper, May 1995, p. 5.)

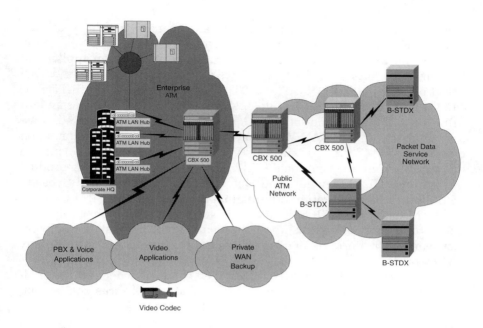

Figure 7.3 Cascade's ATM integration phase 3. (*Source:* Cascade Communications Corporation, "Cascade 500 High Scalability ATM Switch Applications & Technology," Network White Paper, May 1995, p. 5.)

a slot and cost viewpoint—CP and I/O modules, as well as power supplies and fans, can be provisioned redundantly and hot-swapped.

The multiservice heart of the B-STDX is its proprietary control software.

As part of its symmetrical architecture, all switch ports are software-defined as trunks, either network-to-network interface (NNI) or user-to-network interface (UNI). Any port can be used for any of the software-configurable interfaces, whether FR, SMDS, or ATM. (Flash memory and nonvolatile configuration memory are employed in effecting local or remote downloads.) Failed PVCs are dynamically rerouted using the open shortest path first (OSPF) algorithm developed by Cascade's John Moy.

There are a number of other useful features that span the three technologies. One feature, called OPTimum (Open Packet Trunking), allows public packet networks based on FR, SMDS, or ATM to be used as trunk links between switches, desirable for both tariff and availability reasons. Another software component, QuickPath, lets operators choose, regardless of mode (FR, SMDS, ATM), the network QoS that most clearly matches the needs of their applications and manage the delay characteristics of each virtual circuit in the network.

For network management, there is CascadeView/UX, an application built on Hewlett-Packard's OpenView. CascadeView/UX offers a GUI based on Motif and X11R5. The B-STDX has a native SNMP (IETF RFC 1157) agent supporting a Cascade enterprise MIB that includes standard MIB II (RFC 1213) and other FR and ATM MIBs. If desired, CascadeView/UX allows LAN/WAN network management to be combined on a single SNMP platform, in-band via Ethernet or IP or out-of-band via the serial line Internet protocol (SLIP). As the CascadeView/UX product incorporates HP's Openview and Network Node Manager, and Sybase's RDBMS for its configuration information, it requires a Sun SPARC of 256 MB RAM with over 2 GB of hard disk to support a network of 11 or more nodes. It is symptomatic of the complexity involved that the network management platform of an edge switch easily exceeds the cost of many ATM workgroup switches.

Within its B-STDX multiservice edge switch, Cascade has the challenge of staying current with the evolving features of three similar, but ultimately different, high-speed packet technologies.

7.3 B-STDX ATM MODE

Turning first to the B-STDX's ATM implementation, the most unique feature is how Cascade has accommodated ATM's multimedia structure—voice/CBR, real-time video/VBR-RT, nonreal-time video/VBR-NRT, nonreal-time data/ABR/UBR—into its hardware architecture. In its "Quad-Plane Architecture" each kind of traffic is handled by its own dedicated switching plane. Further, each of the two VBR planes are further subdivided into four separate subcategories, providing a total of 10 QoS classes. Supporting this are two levels of buffers, 128K cell buffers partitioned into four programmable planes to support QoS, and port buffers up to 96K per OC-3c module, depending on expected network delays.

Currently the B-STDX software supports ATM's DXI and UNI version 3.0 for AAL 5. Cascade's topology manager, the Virtual Network Navigator (VNN), provides enhanced OSPF with distributed routing tables in each I/O module and a dedicated signaling processor on every line card. Its VNN allows the specification of end-to-end delay, CDV and CLR, administrative path control ("policy routing"), and allows VPNs. Although it transforms the B-STDX into a useful edge product today, they admit that "VNN is a Cascade value-added feature for the interim period while the standards and solutions mature" [1]. Call admission control and traffic management are similarly proprietary via Cascade's Call Master software.

With its B-STDX positioned on the ATM edge, Cascade brings several interesting perspectives to the performance discussion. For one, it sees SVCs as the "currency" of ATM: "Virtual Connection capacity is the truest measure of the revenue-generating capacity of an ATM switch"[2]. With today's traffic largely generated on LANs, LANE, IETF 1577, and MPOA will dominate SVC requests. Further, in the period before ABR gets flow control standards, Cascade sees UBR as today's answer for most data traffic. The company believes that its implementation of UBR with early packet discard (EPD) and fair queuing (FQ), "ensures that UBR connections get throttled proportionately to their transfer rate so that high cell rate users do not monopolize the available bandwidth." The two, EPD and FQ, it sees as "mak[ing] ATM UBR equivalent to existing datagram networks" [3]. Finally, sensitive to the problems of cost and local bandwidth availability, the B-STDX offers an usually broad span of speeds, from 56 Kbps to (future) OC-3c 155 Mbps.

7.4 B-STDX FR MODE

The Cascade B-STDX's bread and butter has been frame relay, and its offering is particularly comprehensive. Cascade's FR implementation offers a huge speed range: from 9.6 Kbps to 45 Mbps, DTE or DCE. It supports the full line of FR Forum PVC/SVC interfaces, FR switch interface (UNI-DCE), FR feeder interface (UNI-DTE), or FR network-to-network interface (NNI), each of which is available on a leased line trunk, multicast, or OPTimum interswitch trunk.

As appropriate to a data switch on the edge, the Cascade B-STDX offers numerous options in the area of protocol conversions:

- If the incoming is X.25/HDLC and SNA/SDLC, it encapsulates it to frame relay format (FRAD).
- If the incoming is point-to-point protocol (PPP), or encapsulated protocols such as IP, Novell NetWare's SPX/IPX, or transparent bridging, it can convert them to FRAD.
- If the incoming is FRAD, it can convert to either ATM's frame relay NNI internetworking standard or use a Cascade OPTimum trunk with ATM.

Although not as comprehensive as ATM's, frame relay allows facilities for relay rate monitoring and enforcement. Cascade's B-STDX includes CIR as well as settings for committed burst size (Bc) and excess burst size (Be). For congestion avoidance, Cascade uses both forward explicit congestion notification (FECN) and backward explicit congestion notification (BECN). Cascade's congestion recovery employs ANSI/ITU standards.

7.5 B-STDX SMDS MODE

Where the FR market is relatively mature and the ATM market growing, SMDS remains a kind of stealth technology. Where ATM basks in high bandwidth and multimedia glitter and FR advertises itself as the most efficient data transfer workhorse, SMDS has not found an applications identity separate from its Regional Bell Operating Company (RBOC) providers. Guided exclusively by Bellcore for most of its technological life, there is now a SMDS Interest Group, akin to the ATM and FR Forums with a similar promotional and standards development agenda. As Cascade B-STDX sales have been heavily to carriers, their SMDS offering heavy reflects RBOC priorities. The Cascade B-STDX offers the following interfaces, each linked with an RBOC role:

- Subscriber-network interface (SNI), which connects end-users;
- SMDS data exchange interface (DXI) by which multiple B-STDX platforms may be connected with OPTimum trunks to build a large, frame-based SMDS switching system;
- SMDS access server functionality per Bellcore TA-1239 specifications, which furnishes local SMDS switching for low speed (less than T1) and for the concentration of traffic for large central offices;
- SMDS intercarrier interface (ICI), which allows other local RBOCS and interchange carriers (IXCs) to interconnect their separate SMDS networks;
- Server-switch interface (SSI), which, with the ICI interfaces and in conjunction with an SMDS CSU/DSU, is used to connecting to other cell-based SMDS networks.

In the above roles, the B-STDX offers speeds from 56 Kbps to 45 Mbps. In the future, SMDS/FR interfaces and support of the FRF.5 network internetworking and FRF.8 service interworking standards is planned.

Viewing the B-STDX at a distance, perched on the network edge, one cannot but be impressed by the huge speed range (9.6 to 155 Mbps) supported as well as the variety of WAN interfaces, with different kinds of DS-3 and E3 options (see Table 7.1).

Table 7.1
Service Support with Cascade's B-STDX 9000 Multiservice WAN Platform

IO Modules	Port Speeds	ATM	SMDS	Frame Relay
Universal IO (V.35, X.21, RS 449)	Up to 8 Mbps	Yes	Yes	Yes
Unchannelized T1/E1	1.54/2.05 Mbps	Yes (DXI)	Yes	Yes
Channelized T1/E1	24 or 30 56/64 Kbps channels	Yes (DXI)	Yes	Yes
ISDN	24 or 30 56/64 Kbps channels	Planned	Yes	Yes
HSSI	up to 45 Mbps	Yes	Yes	Yes
DSX-1	1.54 Mbps	Yes (DXI)	Yes	Yes
T1/E1 circuit emulation over ATM	1.54/2.05 Mbps	Planned	No	No
ATM UNI T1/E1	1.54/2.05 Mbps	Yes	No	No
ATM UNI T3/E3 switching	45/34 Mbps	Yes	No	No
ATM OC-3c UNI/STM-1 switching	155 Mbps	Q4 '95	No	No

Source: Cascade Communications Corporation, "B-STDX 8000 and 9000 Multiservice WAN Platform," 1995.

Nonetheless, there remain conspicuous gaps in popular services now, such as T1 and T3, and even bigger ones to come, such as OC-3c (155 Mbps) and OC-12 (622 Mbps), which will offer opportunities for inverse multiplexing, a new round of carrier-provided fractional services, and VARs selling FR and ATM CIR-based services.

7.6 NEWBRIDGE NETWORKS CORPORATION

Newbridge Networks Corporation, headquartered in Kanata, Ontario, dates from 1986. Its first product was the Mainstreet Bandwidth Manager, a TDM "smart" multiplexer. Today, Newbridge is best known for its strong positions in the T1/E1 multiplexer and FR markets. Estimates by International Data Corporation in 1995 credit Newbridge with 26.2% of the world T1/E1 multiplexer market and 21.1% of the frame relay market [4]. In 1987, Newbridge added its Intelligent Channel Bank, in 1988, its

Intelligent Network Station, and, in 1989, went public. The year 1992 saw the introduction of a frame relay switch, the Packet Transfer Exchange, and in 1993, in conjunction with MPR Teltech, it introduced an ATM edge switch, the 36150. In 1994, Newbridge added to its ATM line with the 36170 Multiservices Switches and VIVID LAN switches. Newbridge's steady stream of products can be seen as something of a parable of today's fast-moving communications equipment market: One need run hard just to stay in place. As ATM equipment can be seen as successor technology to TDM and FR equipment, Newbridge, like Cisco with its routers, must run the risk of cannibalizing its TDM and FR markets with its ATM switches.

Nonetheless, Newbridge brings a number of important advantages to the ATM market. Most important is a worldwide customer base and reputation, particularly with the carriers. Too, it is one of the few ATM vendors that today offer what they call an "overall ATM portfolio"—ATM product offerings spanning NICs to enterprise switches (Figure 7.4).

These products include the following:

- NICs;
- "Ridge" LAN switches:
  Yellow Ridge Ethernet Switch with ATM interface;
  Blue Ridge Token Ring Switch with ATM interface.

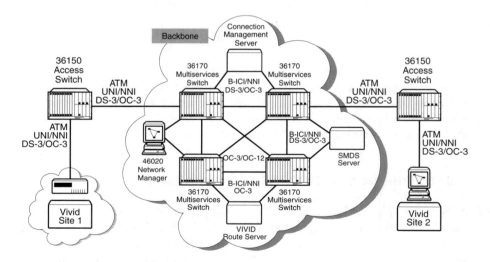

Figure 7.4 Newbridge's long-term, end-to-end network architecture. (*Source:* Ali, I., "Newbridge ATM, The Total Networking Solution," Newbridge Networks Corporation, April 1994, p. 4.)

- VIVID ATM Workgroup Switch;
- ATM Route Server;
- Network Manager, via 4602 Mainstreet Intelligent Network Station TM software on a Sun SPARCstation;
- 36150 Access Switch;
- 36170 Multiservices Switch;
- 36190 Core Services Switch.

Similar to Cascade, Newbridge's most successful product to date, the 36150 Access Switch discussed below, not only is an ATM edge switch but a transition-enabling product, offering FR, SMDS, LAN bridging, and (via ATM's CBR) interoperability between Newbridge's ATM and TDM products. Probably least significant, Newbridge enjoys an early lead in the relatively undeveloped WAN segment of the ATM market, with 52.2% of the ATM WAN market in 1995 [5].

7.7 NEWBRIDGE 36150 ACCESS SWITCH

As noted earlier, the 36150 (Figure 7.5) is a popular access or "edge" switch. As of mid 1995, Newbridge claimed 900 switches and 100 customers globally for the 36150 [6]. The 36150 can serve as LAN bridge, concentrator, switch, or adaptor, and handles ATM, FR, SMDS, and TDM as well as LAN bridging. One of its most interesting features, aimed at the telemedical/distance learning constituencies, is its capabilities for audio/video support. That option supports bidirectional, full-motion, near-National Television Standards Committee (NTSC) quality Joint Photographic Experts Group (JPEG) II video applications. Using ATM's AAL 1, the 36150 supports variable bit rate video compressions at 7.8, 10.2, 12.7, 25.4, 38.2, 42.4, and 127.2 Mbps.

The 36150 switch hardware is fairly straightforward. It consists of a basic nine-slot shelf containing switching cards, power supplies, alarm card, and control cards. All can be duplexed for additional fault tolerance. Universal card slots are employed for transmission and service adaptation cards. The switch itself is expandable, from 4 to 16 ports at up to 155 Mbps each. As befits Newbridge's substantial overseas presence, there are numerous power and mounting options—100, 110, 200 VAC or 48 VDC, and rack, cabinet, and desktop mounts.

In interfaces, Newbridge offers ATM interfaces both local (ATM Forum multimode fiber, single-mode intermediate and long reach 100 Mbps/140 Mbps) and wide area, (SONET OC3c, T3/E3 Bellcore TR-TSY-000499 compatible, physical layer convergence protocol, PLCP, synchronous digital hierarchy, SDH, synchronous transfer mode, STM) [7]. Under

Figure 7.5 Front panel view of Newbridge 36150 Access Switch. (*Source:* Newbridge Networks Corporation, "LAN Interconnectivity Using 36150 Access Switch," 1993.)

the rubric of "service adaptation interfaces," it offers Ethernet, FDDI, compressed NTSC video, TDM, and E1 TDM.

Software for bandwidth and network management has historically been Newbridge's strong suit. Optimizing traffic management in particular is central to the ATM technology. Newbridge calls its "Scaleable Multi-Priority Allocation of Resources and Traffic" or SMART, and it describes its objectives as follows:

"...SMART [is] to maximize nodal and network throughput, and minimize traffic delay, delay variation, and cell discard. The challenge is to handle often incompatible traffic requirements associated with CBR, VBR, ABR, and UBR services. At the same time, the requirements of the existing connections at any given instant have to be fully supported while additional connection requests are being negotiated. Thus, the full implementation of SMART involves the constant monitoring of network and nodal resources and real time serving of traffic requirements" [8].

Several tables below summarize the complexity of these conflicting objectives. The first, Table 7.2, shows the threshold QoS values for ATM service categories for, respectively, cell loss (CLR or cell loss rate), cell delay (CTD or cell transfer delay), and jitter (CDV or cell delay variation).

The second, Table 7.3, shows the bandwidth guidelines to which a given connection within a service category must adhere. Depending on the service category, there are parameters for peak cell rate (PCR), sustainable cell rate (SCR), and minimum cell rate (MCR).

The third, Table 7.4, deals with connection admission control (CAC). Here, one has the relationships between service categories and PCR, virtual bandwidth (Vbw), the link rate, and the MCR.

Table 7.2
Threshold Values for ATM Service Categories

Service Category	CLR $CLP_0\sim$	CTD msec	CDV msec
CBR	1.7–10^{-10}	150	250
VBR-RT	10^{-7}	150	250
VBR-NRT	10^{-7}	Bounded	Unspecified
ABR	Very low	Unspecified	Unspecified
UBR	Engineered	Unspecified	Unspecified

Source: Girous, N., and D. Chiswell, "Traffic Management for ATM LAN Data Networks," Newbridge Networks Corporation, 1995, p. 5.

Table 7.3
Source Traffic Descriptors by Service Category

Service	PCR	SCR	MCR
CBR	×		
VBR	×	×	
ABR	×		×
UBR	×		

Source: Girous, N., and D. Chiswell, "Traffic Management for ATM LAN Data Networks," Newbridge Networks Corporation, 1995, p. 5.

Table 7.4
Common Vbw Values for CAC Implementation

Service Category	Bandwidth Allocated
CBR	$PCR \leq V_{bw} \leq$ link rate
VBR	$SCR \leq V_{bw} \leq PCR$
ABR	$V_{bw} = MCR$
UBR	Engineered

Source: Girous, N., and D. Chiswell, "Traffic Management for ATM LAN Data Networks," Newbridge Networks Corporation, 1995, p. 6.

In network management, Newbridge builds on the success of its 4602 Mainstreet Intelligent Network Station software. The product, which allows network and node monitoring across Newbridge's entire line of switches, muxes, and channel equipment, runs on a Sun workstation. It continues the use of a Newbridge-proprietary messaging format for monitoring; interestingly, only in the LAN area with their VIVID ATM workgroup switch does Newbridge use an SNMP-based system, the VIVID System Manager [9].

Via a graphical user interface (GUI), network operator using the 4602 Network Station software can perform the following connection, fault, and management functions:

- Routing through the switch matrix;
- Controlling bandwidth for virtual circuit connections;
- Configuration of ports, cards, and shelves;
- Performing card diagnostics;
- Locating cards in the system;
- Collecting alarm and status information from cards;
- Resetting cards and taking them in and out of service.

Additionally, the 4602 Intelligent Network Station software enables a number of general management functions:

- Inventory management;
- Connection management;
- Network event and alarm monitoring;
- Test management;
- Traffic monitoring;
- Performance management;
- User environment management, including access delineation, network view customization, and password protection;
- Billing statistics.

Newbridge's network management platform communicates with nodes either locally or remotely via two options. One can employ an RS-232 connection to a local node or, via a "reach through capability," to distant nodes. Or, more typically, one can connect through the Ethernet port on a local node's shelf controller and use virtual channel connections to reach distant nodes.

7.8 SUMMARY

In this chapter, one sees two vendors, one a startup (Cascade), and the other a leading TDM supplier (Newbridge), each packaging ATM for the network edge. The similarity of their approaches—a multiprotocol support strategy in which ATM is but one of several technical tools—is reflective of the complementary roles played by technological "incumbency" (the "legacy networks"), the unevenness of local access to competitively priced high-bandwidth offerings, and a calculated attitude toward the rate of "application push" and the need for ATM technology.

End Notes

[1] Cascade Corporation, "Network White Paper," May 1995, p. 7.

[2] Cascade Corporation, "Network White Paper," May 1995, p. 8.

[3] Cascade Corporation, "Network White Paper," May 1995, p. 9.

[4] Curtis Price, "Data Communications Technology; Preliminary 1994 WAN Equipment Update," International Data Corporation, March 1995, Report #9799; Csenger, M., "In Front of the ATM Curve," *Network World*, Jan. 23, 1995, pp. 65–67.

[5] Curtis Price, "Data Communications Technology; Preliminary 1994 WAN Equipment Update," International Data Corporation, March 1995, Report #9799.

[6] Newbridge News Release of 7/17/95 titled "Newbridge Networks Announces Major New Release of the 36150 Mainstreet ATMnet Access Switch."

[7] "Long reach" refers to optical sections of approximately 25 km or more in length and is applicable to all SONET rates.

[8] Giroux, N., E. Shafik, and M. Gassewitz, "ATM Traffic Management," Newbridge White Paper, March 1995, p. 10.

[9] Ali, I., "Newbridge ATM—The Total Networking Solution," Newbridge White Paper, April 1994, p. 3.

ATM WAN Access Products   8

As noted in the prior chapter, the so-called network edge is not a tidy place today and is likely to remain that way. Clearly, as anticipated by the Cascade and Newbridge offerings, packet, frame, and cell-based technologies will coexist, and the relatively stable environment of voice is likely to remain segregated from the dynamic environment of computer data. (Europe may offer an exception to this observation, where the high costs of leased circuits offer large rewards for multiplex and compression schemes.)

From the vantage of the network edge, this chapter overviews "legacy attachment products" where, whether because of site size or equipment costs or the peculiarities of legacy equipment, the edge will not be served by an ATM switch but by one of several kinds of multiplexers or channel service units (CSUs)/digital service units (DSUs). Although there is considerable functional overlap, there soon will be at least three classes of ATM WAN access products. They are as follows, presented in order of increasing cost and complexity:

- ATM CSUs/DSUs, which provide service termination and enable frame/packet access to ATM WAN media;
- ATM inverse multiplexers (AIM) which will disaggregate and distribute a broadband service stream over several lesser capacity ATM WAN physical links;
- ATM service multiplexers, which concentrate and convert various data protocols into a multiplexed cell stream for transmission onto ATM WAN media.

Common to ATM WAN access products is employment of the ATM Forum's 1993 Data Exchange Interface (DXI), a generic data transfer specification, implementable in software, that lets ATM WAN access products connect to any standard or serial interface, such as the high-speed serial interface (HSSI) or V.35. This allows legacy equipment to use ATM networks without having to wait for equipment providers to field an ATM interface. The practical effect is to allow router and other data terminal equipment (DTE) vendors to inexpensively upgrade their products to make them compatible with DXI-compliant WAN access products. The upgraded routers provide addressing and packetized data for transmission to the WAN access device, which, in the simplest case, converts the packets to cells and assigns them to the proper ATM path.

8.1   ATM CSUs/DSUs

The capabilities of ATM CSUs/DSUs vary by vendor and price, but a number of functions are typical. Many are software settable for either ATM or SMDS, depending on the WAN technology. All perform the channel service unit "service termination" function. All perform ATM (or SMDS) segmentation and reassembly (SAR). All communicate via DXI with the LAN and via UNI with the WAN; some can communicate via UNI with the LAN as well. All provide a rich set of LAN and WAN interfaces and interface adaptations. Most provide support for ATM AALs, 3/4 (SMDS), and AAL 5 (packet data) as well as supporting clear channel operation.

8.2   ATM INVERSE MULTIPLEXERS (AIMs)

The ATM Forum's AIM specification is not expected to be complete until late 1996. Nonetheless, their eventual appearance is a virtual certainty. ATM inverse multiplexers, similar to their T1/E1 cousins, will disaggregate and distribute a broadband service stream over several lesser capacity physical links, thereby reducing costs or compensating for the unavailability of higher bandwidth links. Today's popular channel sizes typically offered by the ex-Regional Bell Operating Companies (RBOCs) and interexchange carriers (IEXs), T1 @ 1.5 Mbps, T2 @ 6.2 Mbps, T3 @ 45 Mbps, OC-1 @ 55 Mbps, OC-3 @ 155 Mbps, are simply too expensive and coarse in granularity for many cost-conscious private network users. The wide popularity of "multirate" or "fractional" T1 and T3 services provides ample testament to this situation. Further, many locations cannot obtain higher bandwidth services like T3, much less the optically based offerings, without paying huge installation costs or committing themselves to expensive, multiyear contracts. So, inverse multiplexers, popular today in the T1/E1 marketplace, will soon be available for ATM, and likely SMDS as well. Probable too, they will perform all of the alphabet soup of CSU/DSU functions: CSU, DSU, SAR, DXI to UNI, AALs 1, 3/4, and 5 support, as well providing the usual LAN/WAN interfaces and interface adaptation.

8.3 ATM SERVICE MULTIPLEXERS

Service multiplexers are concentrators. Multiple steams of data (and often multiple data protocols) enter, typically from a LAN, and emerge as cells to the WAN. At least occasionally, the sources will exceed the capacity of the WAN bandwidth, forcing the service multiplexer into a policing role. The typical objective of an ATM service multiplexer is to mediate the huge speed gaps between legacy services like SNA, TCP/IP, and IPX traffic and new applications like multirate digital videoconferencing, all the while aiming for low delay and high WAN bandwidth utilization. In the future, with a standard way of handling VBR voice transmission, voice services are also likely to become multirate, and voice will become a major contributor to the service multiplex. So, when the reader encounters terms in vendor literature referring to "pooling," "stepup" muxing, or "boundary integration," and the inputs are legacy equipment, some sort of service multiplexing is described or proposed. Figure 8.1 contrasts the roles of DSUs, concentrators, and ATM edge switches in a multiservice network with legacy elements. As typically packaged, the service multiplexer performs as well the usual range of CSU/DSU functions—CSU, DSU, SAR, DXI to UNI, and AAL 1, 3/4, 5—and offers the panoply of LAN/WAN interfaces and adaptations.

Table 8.1 presents a listing of current companies in the access device market. Included is information on products, applications, description, ATM interfaces, throughput, features, interoperability with ATM vendors, and a U.S. price list.

8.4 ADC KENTROX

ADC Kentrox, located in Portland, Oregon, is one of the leaders in providing products in what it calls "carrier services" and "packet services." ADC Kentrox, in turn, is a wholly owned subsidiary of ADC Telecommunications, Inc., of Minneapolis, Minnesota, a supplier of telecommunications equipment. The Oregon subsidiary makes a broad line of DSUs (DataSMART intelligent data service units, IDSU), ATM access concentrators (AAC), ATM-ready Ethernet switches (ATMosphere backbone

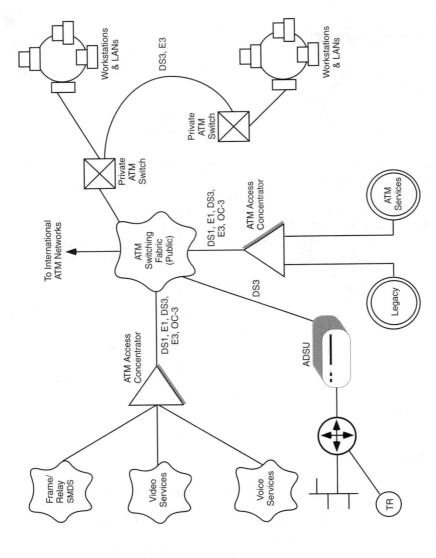

Figure 8.1 ATM access concentration. (*Source:* ADC Kentrox, *Complete Network Access Solutions,* Portland, OR: ADC Kentrox, July 1994, p. 20.)

Table 8.1
ATM Access Devices and Other Interfaces

Vendor	Product(s)	Workgroup	LAN backbone	Private network	Carrier network	Description	1.5–2 Mbps T1/E1	34–45 Mbps T3/E3	100–140 Mbps TAXI	155-Mbps UTP/coax	155-Mbps fiber	622-Mbps fiber	Other
		Applications					ATM Interfaces						
ADC Fibermux Chatsworth, CA	ATMosphere Backbone CBM (Crossbow Backplane Module)	×		×		Module that plugs into vendor's Crossbow hub, provide Ethernet-frame-to-ATM-cell conversion			×				
ADC Kentrox Portland, OR 800-733-5511	ATM Access Concentrator-3 (AAC-3); available 3Q			×	×	Accepts non-ATM data and interfaces (including frame-relay, SMDS, HSS, HDLC) for transport over an ATM network	×	×					
Advanced Telecommunications Modules, Inc. Sunnyvale, CA 408-523-1400	Virata Store; available August	×				Multimedia file-server and storage module (video mail, video-on-demand, up to 8-GB storage capacity) with direct ATM network connection				×	×		Either UTP or fiber interface
Cabletron Systems, Inc. Rochester, NH 603-332-9400	Bridge/Router Interface Module (BRIM)	×				Module that plugs into vendor's hubs, provides LAN-to-ATM conversion (including FDDI); supports one ATM interface			×		×		Modular port interface modules
Cisco Systems, Inc. San Jose, CA 408-526-4000	ATM Interface Processor (AIP); model CX-AIP	×				Module that plugs into vendor's 7000 series router; supports virtual LANs, routing over ATM and adaptation of frame-relay, SMDS over ATM		×	×	×	×		

Table 8.1 (Continued)
ATM Access Devices and Other Interfaces

Vendor	Throughput	Selected Features								Interoperability (with ATM switches)	U.S. Price List
		Supports SVCs	Congestion control	SNMP Management	LAN emulation	CBR traffic	VBR traffic	ABR or UBR traffic	ATM Forum version		
ADC Fibermux Chatsworth, CA	About 100 Mbps (cell and packet rate not specified)		Ltd	×	×		×	×	3.1	None specified	$9,995 for module; hub and chasis required (not included)
ADC Kentrox Portland, OR 800-733-5511	1,170,000 ops; 584,000 pps (approximately 490 Mbps)			×	×		×	×	3.1	PVC interoperability: *FORE*	$23,000 to $60,000, depending on configuration
Advanced Telecommunications Modules, Inc. Sunnyvale, CA 408-523-1400	330,000 cps; 100,000 pps (about 120 Mbps)		×	×		×	×		3.0 & 3.1	Bay Networks, Cisco and *FORE*	$25,000 (includes storage system with 8-GB capacity)
Cableton Systems, Inc. Rochester, NH 603-332-9400	Not specified	×		×					3.0	PVC and SVC interoperability: *Fore*	$3,290 (100-Mbps TAXI interface) to $12,995 (155-Mbps fiber)
Cisco Systems, Inc. San Jose, CA 408-526-4000	375,000 cps; 110,000 ps	×	×	×	×		×		3.0 & 3.1	PVC interoperability: AT&T, *Fore*, General DataComm, GTE, Network Equipment Technology and Newbridge; SVC interoperability	$18,000 to $25,000

Table 8.1 (Continued)

Vendor	Product(s)	Applications				Description	ATM Interfaces						Other
		Workgroup	*LAN backbone*	*Private network*	*Carrier network*		*1.5—2 Mbps T1/E1*	*34—45 Mbps T3/E3*	*100—140 Mbps TAXI*	*155—Mbps UTP/coax*	*155—Mbps fiber*	*622—Mbps fiber*	
CNET U.S.A. San Jose CA 408-954-8000	CN9100	×	×			Ethernet-to-ATM switching hub, does MAC-layer adaptation of 4 (expandable to 6)Ethernet segments to an ATM network				×			Single-mode fiber interface
CrossComm Corp. Marlborough, MA 508-481-4060	ClearPath Edge Router	×	×	×	×	Module that plugs into router, provides LAN (Ethernet and/or token-ring) access to ATM network					×		
Digital Equipment Corp. Littleton, MA 800-457-8211	DECNIS ATMcontroller 631; available August	×	×	×	×	Module that plugs into router, provides multiprotocol routing over an ATM network; 1 or 2 modules per DECNIS 600 router		×			×		Single-or multimode fiber options
	ATM line card for GIGA switch/FDDI system		×	×	×	Module that plugs into FDDI switch, provides bridging between remote FDDI LAN over ATM network: up to 11 ATM interfaces					×		Single-or multimode fiber
Digital Link Corp. Sunnyvale, CA 408-745-6200	W/ATM GateWay Access Multiplexer			×	×	Modular, multislot system that translates, multiplexes and/or switches traffic fromup to 35 T1/E1 user interfaces over 1 or 2 DS-3 ATM interfaces		×					

Table 8.1 (Continued)

Vendor	Throughput	Supports SVCs	Congestion control	SNMP Management	LAN emulation	CBR traffic	VBR traffic	ABR or UBR traffic	ATM Forum version	Interoperability (with ATM switches)	U.S. Price List
CNET U.S.A. San Jose CA 408-954-8000	About 50,000 pps	x					x	x	3.0	PVC interoperability: Telecommunications Laboratories	$10,000
CrossComm Corp. Marlborough, MA 508-481-4060	About 350,000 cps; throughput limited by LAN bandwidth	x		x	x		x	x	3.0 & 3.1	General DataComm and NetEdge	$18,000 to $20,000 for system with 1 ATM and 4 LAN ports
Digital Equipment Corp. Littleton, MA 800-457-8211	180,000 cps; 48,000 pps (about 70 Mbps in each direction)		x	x	Ltd		x	x	3.0	PVC interoperability with *Fore* and General DataComm	$12,500 to $18,000 for ATM module
	170,000 to 260,000 cps, depending on packet size, per 155-Mbps interface; 170,000 pps			x			x	x	3.0	PVC interoperability with *Fore* and Bay Networks	$18,000 to $20,000 per ATM module
Digital Link Corp. Sunnyvale, CA 408-745-6200	192,000 cps					x	x		3.1	None specified	None specified

Table 8.1 (Continued)

Vendor	Product(s)	Workgroup	LAN backbone	Private network	Carrier network	Description	1.5—2 Mbps T1/E1	34—45 Mbps T3/E3	100—140 Mbps TAXI	155-Mbps UTP/coax	155-Mbps fiber	622-Mbps fiber	Other
		Applications				Description	ATM Interfaces						Other
FastComm Communications Corp. Sterling, VA 703-318-7750	Lan SARgent	×	×	×		Stand-alone devices adapt non-ATM interfaces (including Ethernet LAN, T1 or other serial interface) for transport over T1 ATM link	×						
Lannet Data Communications Ltd. Irvine, CA 714-752-6638	LEA Edge Adapter and ILEA Integrated Edge Adapter, available August	×				Stand-alone Ethernet switch and module that plugs into vendor's MultiNet hub					×		
Litton–Fibercom Roanoke, VA 703-342-6700	CAM 7650			×	×	Multiplexer that adapts non-ATM interfaces (including Ethernet, token ring and frame relay) for transport over ATM		×			×		
NetStar, Inc. Minneapolis, MN 612-943-8990	GigaRouter ATM Card	×	×	×	×	Module that plugs into vendor's router, interconnects LANs (including FDDI) and serial interfaces with ATM network					×		
Newbridge Networks, Inc. Herndon, VA 714-758-0100	VIVID Yellow Ridge	×	×	×		12-port Ethernet switch that adapts LAN traffic for transport over ATM					×		
Odetics, Inc. Anaheim, CA 714-750-0100	ATM LIMO Series	×	×	×	×	Series of daughterboards sold as OEM subassemblies that provide an ATM interface for systems; transparent to network layer processing and protocols		×			×		

Table 8.1 (Continued)

Vendor	Throughput	Selected Features								Interoperability (with ATM switches)	U.S. List Price
		Supports SVCs	Congestion control	SNMP Management	LAN emulation	CBR traffic	VBR traffic	ABR or UBR traffic	ATM Forum version		
FastComm Communications Corp. Sterling, VA 703-318-7750	4,250 cps; 2,259 pps (about 1.67 Mbps, based on E1 circuit)	×	×	×	×		×		3.1 & 3.0	Cascade, AT&T and General Data Comm	$4,750 for model with T1 ATM interface, Ethernet and serial ports
Lannet Data Communications Ltd. Irvine, CA 714-752-6638	278,000 cps; 208,000 pps (about 140 Mbps)	×		×	×		×	×	3.1	PVC and SVC interoperability: *Fore*	Not specified; first shipments scheduled for August
Litton-Fibercom Roanoke, VA 703-342-6700	About 149 Mbps (over 155 Mbps ATM link)		×	×	×	×	×	×	3.0	None specified	Varies with configuration
NetStar, Inc. Minneapolis, MN 612-943-8990	35,000 pps (about 130 Mbps)			×			×	×	3.0	PVC interoperability: Cisco/Lightstream, General Data Comm, *Fore* and Bay Networks	$18,000 to $20,000
Newbridge Networks, Inc. Herndon, VA 714-758-0100	About 350,000 cps; 178,000 pps	×	×	×	Ltd		×	×	3.1	None specified	$15,599 (12 Ethernet ports and 1, 155 Mbps ATM interface)
Odetics, Inc. Anaheim, CA 714-750-0100	353,000 cps; 526,000 pps (based on bidirection 155 Mbps ATM interface)								3.0 & 3.1	None specified	$1,150 to $2,660

Table 8.1 (Continued)

Vendor	Product(s)	Applications				Description	ATM Interfaces						Other
		Workgroup	LAN backbone	Private network	Carrier network		1.5–2 Mbps T1/E1	34–45 Mbps T3/E3	100–140 Mbps TAXI	155-Mbps UTP/coax	155-Mbps fiber	622-Mbps fiber	
Premisys Communications Inc. Freemont, CA 510-353-7600	Integrated Multiple Access Communications Server with ATM Server Module	×		×	×	Stand-alone, modular system that enables non-ATM sources (including T1, Ethernet, PBX, video) to connect to an ATM network		×	×		*		
Retix, Inc. Santa Monica, CA 310-828-3400	ATM Interface Card	×	×			Module that plugs into vendor's RouterXchange 7000 router; supports routing over an ATM network			×				
3Com Corp. Santa Clara, CA 408-764-5000	LinkSwitch 2700 and 2700 TLI Telco Switch	×	×	×	×	Ethernet switching device that interconnects 12 Ethernet ports with an ATM network interface		×			×		Single- or multimode fiber
US Networks Inc. Santa Clara, CA 800-777-4526	Access/One GeoLink; and GeoRim (available September)	×	×			Modules for vendor's GeoRim and Access/One hubs; interconnect LANs (including Ethernet, token ring and FDDI) via ATM			×		×		Single- or multimode fiber
Xyplex, Inc. Littleton, MA 508-952-4700	520 ATM 1/0 Module		×	×		Module that plugs into vendor's Network 9000 hub, interconnects Ethernet segments to an ATM network		×	×		×		*Available later in 1995

Table 8.1 (Continued)

Vendor	Throughput	Selected Features								Interoperability (with other ATM switches)	U.S. List Price
		Supports SVCs	Congestion control	SNMP Management	LAN emulation	CBR traffic	VBR traffic	ABR or UBR traffic	ATM Forum version		
Premisys Communications Inc. Freemont, CA 510-353-7600	73,000 to 95,000 ops; about 40 Mbps throughput (based on DS-3 ATM interface)	×		×		×	×	×	3.0	PVC interoperability: *Fore*	$15,550 per ATM module; base multiplexer system required (not included)
Retix, Inc. Santa Monica, CA 310-828-3400	20,000 cps; 35,000 pps (about 65 Mbps)				×				3.0	PVC interoperability: *Fore*, Bay Networks and Newbridge; SVC interoperability: Network Equipment Technology	$10,000
3Com Corp. Santa Clara, CA 408-764-5000	About 430 Mbps aggregate system bandwidth	×		×	×			×	3.1	PVC and SVC interoperability: *Fore*, Cisco and DEC	$8,400 to $9,600 for system with 1, 155 Mbps or DS-3 ATM interface and 12 Ethernet ports
US Networks Inc. Santa Clara, CA 800-777-4526	49,000 (Access/One GeoLink) to 80,000 (GeoRim GeoLink) pps	×		×	×		×	×	3.0 or 3.1	None specified	$7,000 for GeoRim module; $13,000 for Access/One GeoLink
Xyplex, Inc. Littleton, MA 508-952-4700	Up to 40,000 pps	×	Ltd	×		×	×	×	3.0	PVC interoperability: *Fore*, Cisco, Bay Networks, DEC, Newbridge, Xylan, Fujitsu	$8,995 for module

Source: Mier, E. E., "Building A New Order," *Communications Week*, June 12, 1995, pp. 81, 84.

virtual networking engine) and SNMP management products (the Network Essential SNMP/TELNET manager). For a neat product summary, see Figure 8.2.

A sign of the increasing role of ATM access products was ADC Kentrox's 1995 technology exchange with FORE Corporation, the current leader in the ATM switch market [1]. Under the deal, FORE has the right to use Kentrox's ATM adaptation and concentration product technology. This resulted in FORE announcing enhanced versions of the Kentrox products, dubbed CellPath, in 1996 [2]. For its part, ADC Kentrox was allowed to license FORE's ForeThought internetworking software for inclusion in its line of ATM access concentrators. The chief beneficiary on the deal has not been immediately evident. Clearly, ADC Kentrox will increase box sales in the short run. The risk is that FORE may either prove the better marketer or incorporate ADC Kentrox's access technology directly into its switching product line.

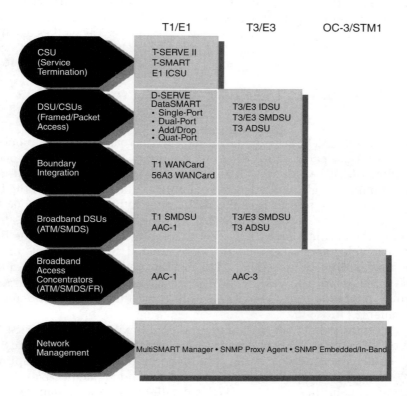

	T1/E1	T3/E3	OC-3/STM1
CSU (Service Termination)	T-SERVE II T-SMART E1 ICSU		
DSU/CSUs (Framed/Packet Access)	D-SERVE DataSMART • Single-Port • Dual-Port • Add/Drop • Quat-Port	T3/E3 IDSU T3/E3 SMDSU T3 ADSU	
Boundary Integration	T1 WANCard 56A3 WANCard		
Broadband DSUs (ATM/SMDS)	T1 SMDSU AAC-1	T3/E3 SMDSU T3 ADSU	
Broadband Access Concentrators (ATM/SMDS/FR)	AAC-1	AAC-3	
Network Management	MultiSMART Manager • SNMP Proxy Agent • SNMP Embedded/In-Band		

Figure 8.2 ADC Kentrox's access products. (*Source:* ADC Kentrox, *Network Profiles*, Portland, OR: ADC Kentrox, 1994, p. 53.)

8.5 ADC KENTROX DATASMART T3 ATM DATA SERVICE UNIT (ADSU)

The ADC Kentrox DataSMART T3 ATM ADSU shows how software—the "DataSMART" element—allows a host of options for connection via ATM, SMDS, or clear-channel mode with DS-3 bandwidth. The sales pitch for the ADC Kentrox access equipment is to maximize one's ATM/SMDS investment. On the DTE side, when in DXI mode, the ADSU typically segments packets received from a router into cells for the ATM network (the SAR function). It supports a variety of DXIs, including ATM versions 1A and 1B, FR, SMDS, and Cisco's version of SMDS. DTE-side physical interfaces include support of HSSI or a number of V.35-based "selectables." Network-side, the T3 network interface can be set in ATM UNI, SMDS SNI, or clear-channel modes, with ATM mode supporting AALs 3/4 and 5. There are the usual options for Bayonet-Neill-Concelman (BNC) connectors as well as alternating current (ac) and direct current (dc) power options. SNMP management is embedded and includes various traps (link up, link down, cold start, user defined) as well as terminal access via Telnet.

8.6 ADC KENTROX T1/E1 ATM ACCESS CONCENTRATOR (AAC-1)

ADC Kentrox's ATM access concentrator (AAC) "family" (a T3 model is expected) uses "cell multiplexing to concentrate multiple data protocols, including SMDS, ATM, and FR as well as isochronous (constant bit-rate) sources onto a single network access facility" [3]. It is telling that the first family member to appear is at the T1/E1 rates rather than the sexier and more costly T3, OC-1, or OC-3 rates. Too, the product is opportunistic in its aim—the current, large FR market to transfer computer data, the BOC and post telegraph telephone (PTT)-dominated SMDS market, or the emerging multimedia-oriented ATM market. Overall, the ADC Kentrox pitch is conservative, with the AAC-1 designed to provide a natural, or gradual, migration to ATM by protecting "investment in embedded-base equipment" [4].

With the AAC-1, all this configurational flexibility has its price: the number of network, DTE, and power source options are huge (Figure 8.3).

Rather than indulge in the combinatorial bliss of a zillion options, one is advised to keep a higher level—and two-sided perspective—of the AAC-1's capabilities. On the DTE side, the AAC-1 can handle ATM sources in either DXI or UNI formats. There are a variety of modules that allow the AAC-1 to support DTE-side equipment at either full (native)

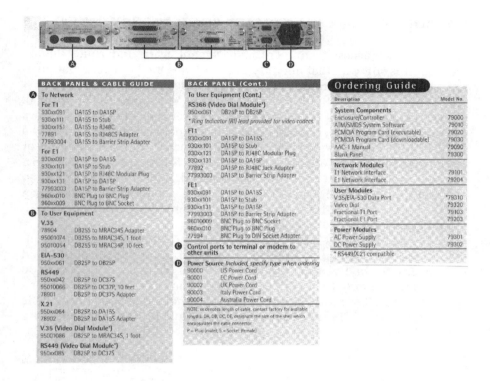

Figure 8.3 Options with ADC Kentrox's T1/E1 ATM access concentrator's back panel and cable ordering guide. (*Source:* ADC Kentrox, "T1/E1 ATM Access Concentrators," August 1996, unpaged.)

rates or at user-specified selectable rates in 8-Kbps multiples (up to 2 Mbps). This allows the network administrator to essentially impose upper boundary traffic management on applications (many ATM NICs allow this feature as well). As expected, there are a large number of DTE-side physical connectors, including CCITT V.35 or X.21 interfaces and a variety of EIA RS-422 options. On the network side, the AAC-1 presents a UNI appearance and supports ATM AALs 1, 3/4, and 5. Again, there are a huge number of WAN connector options, both DA15P- and BNC-based.

Additionally, there are a number of not so obvious benefits in employing the ADC Kentrox AAC-1. It allows support of multiple groups and multiple E.164 addresses per physical link. It provides a network administrator with another source of statistical data on the cell payload. And it allows one to install more than one connection to the ATM (or SMDS) network per physical device, at least potentially making the link more cost effective.

8.7 ADC KENTROX MULTISMART SNMP AGENCY (MSA)

ADC Kentrox's access equipment management again underlines the axiom that it is software that not only differentiates the hardware, but justifies vendor price differences. The ADC Kentrox management software, which runs on top of SunNet Manager, Open View, and NetView/6000, allows remote control of the access devices over the network and allows their integration into a corporate, SNMP-based network management. The Sun/HP/IBM managers can, in turn, be connected to either Ethernet or token ring networks. Called the MultiSMART SNMP agent (MSA), MSA controls the devices in-band over the network and allows, via a daisy-chaining feature, multiple devices at a remote site to be controlled by a single control port. A full-featured schema employing MSA appears in Figure 8.4.

The ADC Kentrox MSA network management is designed for unattended operation and allows one to access alarms, performance data, and diagnostics, as well as configuration data. Via daisy chaining, up to 30 devices are supported per connection by the use of either Cisco router's IP- addressable CPU auxiliary port or a terminal server's IP-addressable asynchronous port. Telnet sessions from the Sun/HP/IBM managers use a menu-like format.

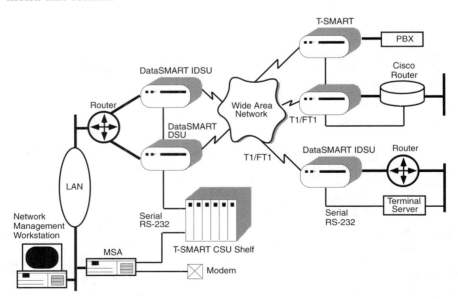

Figure 8.4   ADC Kentrox's MultiSMART SNMP agent (MSA) employed in a hypothetical multiservice network. (*Source:* ADC Kentrox, *Complete Network Access Solutions*, Portland, OR: ADC Kentrox, July 1994, p. 24.)

SNMP traps are alarms that are forwarded by the MSA to the SNMP "trap-host(s)" and include link up, link down, and cold start generic traps as well as user-defined traps. Each trap message contains the device address, name and type of alarm.

Direct device control is available from any workstation on the LAN. One connects via a Telnet session through the MSA to any of the managed WAN access devices. By way of the Telnet session, line status, diagnostics, and administrative tasks can be performed online. Full English menus are provided for the ADC Kentrox devices so that one does not have to interpret the management information base (MIB) variables. MSA supports the current DS1 MIB allowing information on T1 line status, performance, and configuration. T1 line performance can also be polled from the SNMP management application on demand or at regular intervals.

8.8 DIGITAL LINK CORPORATION

Digital Link Corporation is a Sunnyvale, California-based startup whose stated ambition is to be the "leading supplier of easy-to-manage WAN access solutions" [5]. Founded in May 1985 by Vinita Gupta, its current CEO, it became profitable within its first six months. Starting with WAN access products for T1/E1, Digital Link moved on to access products for frame relay and an SNMP manager. It then followed the bandwidth and technology spiral with products for T3/E3, SMDS and ATM, SONET, and Internet access.

As will become evident in the design of their products—particularly the defining role of software—they hope to ride not one, not two, but three WAN technologies. Dataquest's service revenue projections for frame relay are for $2.4B by 1998. By the same date, the Vertical Systems Group projects ATM revenues to reach $339M. SMDS, strong in Europe, is similarly projected by International Data Corporation to reach $153M by 1998 [6]. Following the numbers, Digital Link keeps a foot in three boats (SMDS member, ATM Forum member, Frame Relay Forum member). Also, with attention to the rapidly deregulating and expanding international markets, Digital Link products bear an ISO 9001 Certified Quality Management System seal.

Today, Digital Link offers a continuum of products for both broadband access (T1/E1, T3/E3, OC-1 & 3) and frame and cell packaging (FR, SMDS, ATM). By their own description, they offer "scaleable solutions" for WAN access, from 56 Kbps to 155 Mbps, in three categories—DSUs/CSUs, access multiplexers, and inverse multiplexers—with SNMP network management support. Distribution today is through value-added retailers (VARs), public service providers, and telephone companies. Finally, as a portent of what may be in store, Digital Link recently announced an edge switch, the A/ATM Gateway ATM cell multiplexer, that is based on a "native ATM bus design."

8.9   DL3200 SMDS/ATM DIGITAL SERVICE INTERFACE

Digital Link touts their 3200 SMDS/ATM digital service interface as "the industry standard for ATM and SMDS T3 access" [7]. Combinatorily, its RISC-based hardware base enables a single WAN port in two sizes (DS-3, E3), two DTE interfaces choices (V.35, HSSI), and comes in three software-determined models (DL3200-A, S, or AS) depending on the application environment. It also has the capability of accepting data packets in the high-level data link control (HDLC) format and encapsulating it for SMDS or ATM WAN transmission.

The DL3200A converts frame relay to and from the LAN to ATM for the WAN. It accepts data from frame-relay-compatible DTE (V.35, HSSI) and converts from the frame relay packet address (the data link connection identifier or DLCI) to the virtual path/virtual channel identifier (VPI/VCI) of the ATM network. The data packet is then encapsulated in AAL 3/4, converted to cells, and encapsulated using the DS-3 physical layer convergence protocol (PLCP) for ATM transport.

The DL3200S converts SMDS DXI to and from the LAN to SMDS on the WAN. The 3200S passes data from internetworking equipment (routers, bridges, channel extenders) and converts it for transmission on cell-based networks. More specifically, it receives L3_PDUs over the SMDS DXI via V.35/HSSI. It then segments them into cells of equal length (53-octet L2_PDU cells) with additional overhead for flow and link management. Software is selectable for SMDS access classes 1–5. The resulting data is then mapped into a DS-3 signal via the DS-3's PLCP.

The 3200AS model converts SMDS DXI to and from the LAN to ATM on the WAN. It takes SMDS DXI as DTE and converts the L3_PDUs to ATM cells. The address conversion is from an E.164 (SMDS) address to ATM VPI/VCI.

A typical DL3200 application appears in Figure 8.5.

Digital Link initially offered its own Digital Link network management system (DLMS), a UNIX-based multitasking software package running on a Sun SparcStation as an SNMP manager, and it is still available for the older products. Digital Link's newer products, including those described here, feature integrated ("embedded") SNMP and run under SunNet/OpenView network managers.

Basically, there are four ways to baby-sit the box: direct connection to a SMDS network, dial access via a serial link interface protocol (SLIP)

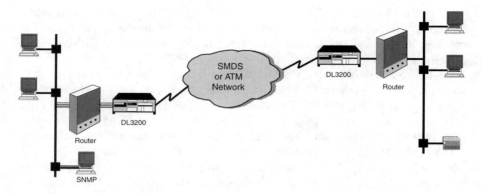

Figure 8.5 A typical DL3200 application environment (*Source:* http://www.dl.com/).

attachment for ATM, asynchronous terminal access, and the front panel display.

Clearly the most desirable monitoring mode—direct connection—is available only for SMDS today. The SMNP agent is accessed in-band with messages carried by the SMDS network with the manager directly connected to the network. An external, non-network connection is not required. With ATM, SNMP connectivity is achieved via dial-up modem and SLIP—a much less desirable state of affairs, particularly if the SNMP managers are not distributed. The built-in asynchronous communications port provides direct terminal access for local technician maintenance and test functions and is the usual manner to effect field upgrades of the software. Finally, the DL3200 has a 16 character front-panel fluorescent display that provides local access to network performance information, unit options, and diagnostic tests.

8.10 W/ATM PREMISEWAY ATM SERVICE ACCESS MULTIPLEXER

Less its new edge switch, Digital Link's W/ATM PremiseWay ATM service access multiplexer represents the company's top of the line. Its function is to integrate data, voice, and video and other customer premises equipment (CPE) devices onto public ATM WAN services and private ATM WAN backbones. Accordingly, Digital Link modestly describes it as the "on-ramp to the Information Highway" [8]. (It should be noted, however, that it split the DSU/service mux category by offering a low-end, two-into-one (dual port) device, with dynamic channel sharing, the DL3202 digital cell multiplexer.)

Digital Link's W/ATM PremiseWay ATM service access multiplexer employs a five-slot modular architecture. There is one network interface module (NIM) and up to four subscriber interface modules (SIMs), each offering either a V.35 or HSSI connection, per unit.

There are two kinds of SIMs: VBR and CBR. The VBR SIM has software-selectable V.35 or HSSI electrical interfaces. A V.35 port supports speeds to 10 Mbps; the HSSI port supports speeds to 50 Mbps. The VBR SIM provides protocol support for FR, SMDS DXI- or ATM DXI-compatible routers, bridges, and channel extension equipment and supports ATM AALs 3/4 and 5. The CBR SIM provides three "structured" (N × 56 Kbps or N × 64 Kbps) or "unstructured" (full T1, 1.544 Mbps) T1/E1 interfaces and a single V.35 interface for delay-sensitive traffic. The CBR SIM provides transparent transport of T1/E1 or fractional T1/E1 (N × 64 Kbps) traffic from PBXs, channel banks, and T1 muxes over ATM via AAL 1 circuit emulation. Last, the CBR SIM supports via transparent transport a V.35 circuit AAL 1 for use with video codecs.

For details of NIMs/SIMs specifications, see Figure 8.6. As these specifications are important to understand and necessary to the support of various options, but also somewhat cryptic, several aids are provided below. The "ppm" in the "Line Code" categories stands for "pulse position modulation." In the "Transmit Distance" categories, "FC-PC" refers to NTT's "face contact and point contact," fiber connectors that feature a moveable, antirotation key allowing good repeatable performance despite numerous matings; "DIN" refers to the German Institute for Standardization. Transmit "LBO" refers to (electrical) "line build out." Under the Adaptation Layers" category, "ATM DXI mode 1a" refers to one of two ATM DXI modes, the other being 1b. With the "Connector Types" category, "Amplimite" refers to a popular connector product of AMP Corporation's Interconnection components and Assemblies Products (ICAP) unit.

On the network side, NIMs offer several options for WAN access. At what might be called (tongue in cheek) the "low end," there is DS-3 or E3. At the high end, the current maximum is either OC-3 ATM or synchronous transfer mode (STM)-1.

An access multiplexer has several tricks to boost goodput and protect itself. Relative to the former, the PremisWay can optionally perform traffic shaping per ATM Forum UNI 3.1. When traffic shaping is enabled, bursts of cells on a virtual circuit are spaced evenly on the output stream according to the peak cell rate. By spacing the cells, the receiving ATM switch is less likely to tag or discard them, thus resulting in better WAN performance. Relative to protecting itself against large bursts of data traffic, the PremisWay can implement intelligent, frame-based discard. When a circuit violates its peak cells rate, the PremisWay allows the current frame to complete transmission, but then discards the subsequent

SYSTEM SPECIFICATIONS

Chassis	Five-slot modular design
	One NIM
	Up to four SIMs
NIM	DS3, E3, OC3/STM-1
VBR SIM	HSSI/V.35 (software selectable)
CBR SIM	Three T1 and a V.35 or three E1
	and a V.35
Traffic Shaping	Per ATM UNI 3.1 leaky bucket
	algorithm
OA&M Support	Loopback, Performance
	Monitoring and Continuity Check
Management	Front Panel, Local
	Terminal, LMI or via SNMP
SNMP	Embedded SNMP agent supports RFC
	1407, RFC 1483 and Enterprise MIB
Diagnostics	Loopbacks, Traffic Generation and
	OA&M cells

NETWORK INTERFACE MODULES (ONE PER UNIT)

OC3/STM-1 OPTICAL NETWORK

Line Rate	
Line Code	1.55.52 Mbit/s±20 ppm
Framing	NRZ (Per G.957)
Physical Media	SONET (Per G.709)
Connector Type	Single Mode Optical
Transmit Distance	FC-PC
Synchronization	Up to 15 Km
	Internal, Network or External

STM-1 ELECTRICAL NETWORK

Line Rate	
Line Code	1.55.52 Mbit/s±20 ppm
Framing	Code Mark Inversion (Per G.703)
Physical Media	SONET (Per G.709)
Connector Type	75Ω coax
Transmit Distance	DIN 47295
Synchronization	Up to 200 M
	Internal, Network or External

DS3 NETWORK

Line Rate	
Line Code	44.736 Mbit/s±20 ppm
Framing	B3ZS
Impedance	Standard or C-bit
Connector Type	75 Ohms Resistive
Pulse Shape	BNC
Transmit LBO	Meets Bellcore TR-TSY-000009
Synchronization	0–225 ft., 225–450 ft.
	Internal, Network or External

E3 NETWORK

Line Rate	
Line Code	34.368 Mbit/s±20 ppm
Framing/Mapping	HDB3
Connector	G832
Transmit Timing	BNC or DIN
	Internal, Network or External

SUBSCRIBER INTERFACE MODULES (UP TO FOUR PER UNIT)

VBR Module

	One per module
Number of Ports	AAL5 and AAL3/4
Adaptation Layers	ATM DXI mode 1a
Protocols Supported	SMDS DXI 3.2
	SMDS Alt. 1.2 (Cisco Interface)
	Frame Relay Two- and Four-Byte Header
Physical Interfaces	V.35 and HSSI (software selectable)
Connector Types	DB25 Female (V.35)
	50-pin Amplimite (HSSI)

CBR Module

Number of Ports	Four per module
Adaptation Layers	AAL1
Physical Layers	Three T1 ports, one V.35 port
Connector Types	DB15 Female (T1)
	DB25 Female (V.35)
T1 Interface Rate	Unstructured (full T1, 1.544 Mb/s)
	Structured (Nx56 Kbp/s or Nx64 Kbp/s)
V.35 Interface Rate	Up to 10 Mbit/s

PHYSICAL/ENVIRONMENTAL

Width	17.2" (43.7 cm)
Height	5.2" (13.3 cm)
Depth	14.0" (35.6 cm)
Mounting	19" or 23" rack
Operating Temp.	0° to 50° C ambient
Storage Temp.	-35° to 85° C
Relative Humidity	0 to 95% Non-Condensing
Altitude	-200 ft. (-60.98 meters) to
	10,000 ft. (3048 meters) ASL

REGULATORY

Complaint	UL1950, CUL, IEC 801,2,3 and 4
	VCCI Class2, FCC 15 Class A

Figure 8.6 Specifications of Digital Link's PremiseWay ATM service access multiplexer. (*Source:* Digital Link Corporation, "PremiseWay ATM Service Access Multiplexer," 1995.)

frame or frames necessary to maintain the desired peak cell rate. This mechanism is both self-protective and reduces congestion-causing retransmissions.

Given the steep costs for recurring bandwidth in the 45/155 Mbps range, effective use and reliable management of that bandwidth is both a high priority and potential product differentiator. Accordingly, the PremisWay offers a number of advanced features for unobtrusive diagnostics as well as a large number of modes by which to effect local or remote equipment management.

In the diagnostic area, the PremisWay supports OAM (F1-F5) flows, including "keep alive" and performance monitoring. This allows the user to check system integrity without disrupting payload traffic, enormously important for what is essentially a funnel to a large pipe. Further, to support testing, the PremisWay also includes a built-in ATM traffic generator that can send cells on any VPI/VCI. This is performed by segmenting a user-defined FR or SMDS frame into ATM cells. The network technician can set the frame length, VPI/VCI, and cell transfer rate. The PremisWay can then send these cells to the ATM network, allowing network staff to test network capacity and performance.

For SNMP-based network management, the PremiseWay's preferred approach is embedded SNMP management and in-band transport in conjunction with either a SunSoft's SunNet Manager or HP's OpenView. The PremiseWay's SNMP agent supports ATM, DS-1/DS-3, E1/E3, and SONET/SDH MIBs as well as an enterprise MIB for specific unit provisioning and tests.

It is in the number of possible management access modes to the access multiplexer(s) where it gets complicated. First, there is the method already mentioned, in-band access through the ATM network. A second way is via PremiseWay support of the local management interface (LMI) for communications to and from DTE devices, allowing a router or other DTE device to query the unit for statistics, alarms, status, and configuration. A third path is via a direct SLIP (serial link interface protocol) attachment. A fourth involves a terminal user interface that allows the unit to be managed by a VT-100-compatible terminal. Fifth is direct use of the unit's front panel display. The first four of these modes could be used, though not with equal ease, to effect code downloads, allowing the addition of new features and capabilities.

8.11 SUMMARY

Even more than in the ATM switch market, the access market will belong to the swift, the first fielding important functional and management features. It is the kind fast-moving, small box, software-intensive area that could be barely recognizable in two years. Accordingly, some prognostications are in order:

- Because of market conditions already noted, ATM inverse multiplexers (AIM) are badly needed and an AIM standard will jump-start that market.

- Competition from FORE, and potentially others, in the DSU market with built-in DSUs is highly likely—think of them as built-in modems à la laptops.
- DSUs sales (whatever the form) will be a monotonic function of edge and WAN switch sales; access multiplexer sales will be more complex, a function of the frequency with which several computer data streams are combined to run over ATM networks or in the muxing in of video and voice data.
- With more "ATM-ready" products appearing, less ATM DXI and more ATM interfaces are likely two years hence.
- There is likely to be increased competition between what could be characterized as fancy access concentrators and cheap edge switches.
- In the longer run, particularly with a standard VBR-RT mechanism for voice, a move away from today's ATM, which could be characterized as a more or less exclusive focus on data networking. In this direction, FORE's new CellPath 200 access concentrator is anticipatory. It offers applications modules, in both four- and eight-port versions, for two kinds of voice modules. The 4WE&M allows PBXs to communicate as peers, and the 2WFXS/FXO provides a local dial tone in a remote location.

End Notes

[1] Petrosky, M., et al., "Fore Announced Switches, Updates Architecture," The Burton Group, Oct. 18, 1995.
[2] Petrosky, M., "ATM Market Maturing," The Burton Group, v1, May 1996.
[3] ADC Kentrox, *Network Profiles*, Portland, OR: ADC Kentrox, 1994, p. 52.
[4] *Ibid.*, p. 21.
[5] Digital Link Corporation, "Company Overview," http://www.dl.com/.
[6] Digital Link Corporation, "The Market," http://www.dl.com/.
[7] Digital Link Corp., "DL 3200 Digital Service Interface," Sunnyvale, CA: Digital Link Corporation, Jan. 1995, p. 1.
[8] Digital Link Corporation, "PremiseWay, ATM Service Access Multiplexer," Sunnyvale, CA: Digital Link Corporation, Sept. 1995, p. 2.

Large ATM Switches   9

Today's large switch market in ATM engenders some honest confusion. As both the number of players and the volumes shipped are quite small thus far, there is a Janus-quality to the promotional literature. One face courts the dedicated private network buyer and describes switches for the core of large "enterprise" networks. The other woos the carrier market and pitches products for the central office (CO). They are, of course, give or take some hardware and software options, the identical product.

One path from this convenient confusion is to describe the large ATM switches functionally in terms of their capabilities. As a group, large ATM switches usually display the capabilities (or limitations) below:

- Offer huge aggregate bandwidth (tens of gigabits) capable of incremental deployment (scalability);
- Support optical carrier (OC)-n and STMn media only;
- Provide redundant (1 + 1) provisioning for fault tolerance on big-bandwidth bundles;
- Claim "carrier-grade" reliability (redundant everything);
- Process ATM only, no legacy protocols;
- Support both UNI/NNI "views;"
- Offer software packages to perform user accounting and chargeback;
- Represent big ticket items with relatively few vendors (Nortel, AT&T, Fujitsu, Alcatel, NEC, Siemens);
- Are available with a choice of network management systems, one proprietary, as well as one based on either the CMIP or SNMP standard;
- Compared to edge switches, contain less total software lines of code (LOC) with a slower release rate.

Further, there is an air of unreality when dealing with the applications that these large switches will be handling. It is not evident that any of the public carriers offering ATM services, such as MFS Datanet, have reached volumes where core switch capacity is an issue; for them, the crucial issue is obtaining a chargeback capability. For the private networks, there are a number of large-scale demonstration networks in the field, many combining distance learning, the exchange of radiological or cardiological images, or digital video-on-demand systems, but it is not clear that these applications will pay their way once the feasibility trials have been completed. The technological metaphors are troubling: Are they battleships in a PT boat war? Are they technological capability in search of an application? Are common carrier costs for big-bandwidth bundles so high as to preclude all but cost-insensitive (military and intelligence) applications?

9.1 NORTEL AND THE MAGELLAN FAMILY

Northern Telecom (NT) Ltd., of Mississauga, Ontario, also known as Nortel, is a $9 billion telecommunications vendor with a long history in voice and packet switching. It spends more than $1 billion annually in R&D and its Bell Northern Research (BNR) laboratory has long been a respected innovator in high technology. Where much of the action in ATM has been led by small startup companies, it was Nortel, along with Adaptive Corporation (part of Network Equipment Technologies, Inc.), Cisco, and StrataCom Inc. (since April 1996, part of Cisco), that were the original founders of the ATM Forum in October of 1991 [1]. Early on, Nortel saw the long-term strategic importance of ATM.

The Nortel response has been the Magellan "portfolio" or "family" of scaleable products to support the broadband multimedia market. Variously described either narrowly as the ATM switches (Passport, Vector and Concorde) or broadly to include the DPN-100 data switch, the network management system (NMS), the FiberWorld portfolio of SONET transport equipment, the Cornerstone portfolio of community network products, and video-on-demand residential products, it is one of the most comprehensive ATM product lines offered by any vendor. Briefly, the lineup, is as follows:

- Magellan Passport is a 1.6-Gbps ATM/frame relay switch positioned as a customer premise equipment (CPE) ATM enterprise network switch or CO ATM multiservice access switch providing ATM services, frame relay, interLAN switching, and intelligent circuit emulation services; it incorporates LAN interface and routing technology from Network Systems Corporation (NSC).

- Magellan Vector is a 2.5- to 10-Gbps ATM network equipment building standard (NEBS)-compliant network access concentration and multiplexing switch that serves as an end node for a mix of data traffic types, including X.25, SNA, token ring, and Ethernet; it includes routing technology from Cisco.
- Magellan Concord is a CO high-capacity, 40-Gbps backbone ATM switch, scaleable to 80 Gbps, designed for use in the core of an ATM network or as part of a large video-on-demand system.

Common to the Magellan line are a unified set of technical capabilities and a philosophy as to where they are distributed. Magellan switches offer a huge number of size, configuration, software, hardware, software, and interface options. Among the common features (discussed more fully later on) are Nortel's proprietary AAL for VBR voice and its multiple priority system (MPS) to deliver QoS and traffic and service management. At the same time, Nortel has made a major effort to stay as standard as possible with other ATM Forum, IETF, ITU-T, ANSI, SONET, and SDH standards. From a network philosophy perspective, Nortel believes that policing, traffic isolation, fairness, shaping, and ABR virtual source/destination support are all best handled at the edge of the service provider's network. As a consequence, Nortel's ATM edge switch, the Passport, has the highest (and most volatile) software content.

Nortel has followed FORE's lead in realizing that early ATM adopters are willing to pay for demonstrable functionality regardless of standards status. Key to competing in this race is skill at cross-licensing, and several of the Nortel relationships have already been noted. Realizing FORE's dominance in the LAN market, Nortel licensed FORE's Fore-Thought ATM SVC routing and ATM LAN emulation technology to provide customers with LAN-to-WAN service compatibility with the leading local ATM switch and network interface card (NIC) supplier [2]. Noted earlier was the use of NSC's routing and bridging software used in Passport's interLAN switching. On the carrier side, Nortel used its Vector switch in teaming with FORE, Sprint, and NEC as part of Sprint's ATM offering [3].

9.2 VOICE NETWORKING

Nortel's use of ATM for voice networking is one of the most effective uses of the ATM technology today. In most corporate networks that carry voice and data, voice still accounts for the lion's share of the bandwidth. Nortel's approach provides significant cost savings through a combination of logical multiplexing, silence detection, and dynamic compression. The overall network cost savings can be considerable. MFS Datanet uses the Magellan Passport switch to offer voice services to its customers and claims use of the technology can lower the cost of voice calls by 25% over carrier tariffs [4].

Nortel shuns the usual ATM AAL 1 circuit emulation approach for voice traffic. Instead, it delivers a high-quality, efficient voice capability by using a proprietary AAL that defines a VBR-RT traffic type while employing several well-known digital voice processing techniques. They include the following:

- *Silence detection and suppression.* When there is no call up on a trunk, bandwidth is available. As voice traffic is typically 50% silence, bandwidth is available in both directions. And when there is a call up and only one person talking, bandwidth to support traffic in the opposite direction is available. In tests, Nortel has found that silence suppression (at a 42% speech activity factor), adjusted by cell overhead (13%) and Q.931 ISDN signaling link (ISL) traffic (5%), yields average gains of 40%.
- *Compression.* A number of schemes can be configured. It can be applied staticly at 32, 24, or 16 Kbps (ADPCM), or, at the cost of slightly diminished voice quality, applied adaptively based on congestion in the network. Performed at the network edge by the Magellan Passport, various feedback mechanisms are employed to notify the traffic source that congestion is pending. The source then "downspeeds" or increases voice compression (where permitted) to alleviate congestion until the congestion is relieved. In the future, Nortel promises even greater efficiencies using state-of-the-art algorithms like International Telecommunications Union (ITU)-8 and low delay-code excited linear prediction (LD-CELP).

Further, Nortel's use of silence suppression and voice compression can be manipulated in a number of ways. Both can be set to operate for all voice calls, or they can be activated when a certain level of network congestion occurs. Net managers also have the option of using suppression and compression for some users while allowing others to use the full amount of bandwidth required for normal voice calls. Also, the Passport edge switches can be set to filter out and discard the signaling transmissions used in conventional voice networks.

Although the cost savings come from the logical muxing of voice streams via ATM and the use of compression and silence detection and suppression, there is a good deal more than meets the eye. Voice quality can be adversely affected by many factors, such as delay variances, echoes, speed clipping, congestion, and cell loss. In particular, relative to speech and silence detection, attack time, speech threshold, and hang-time all affect speech quality. So, to maintain or enhance user-perceived voice quality and minimize potential voice-quality impairments, considerable engineering effort elsewhere was necessary.

One of the complementing capabilities is what Nortel calls a "virtual tandem PBX," the creation of direct, single-hop call routing from PBX to PBX without a large transit (or tandem) PBX. Rather, the network assumes the role. But for the network to be effective as a transit PBX, it must work intelligently with the D-channel call channel signaling.

The Magellan Passport's hybrid architecture allows networking trunks to use either standard ATM trunking or Nortel's proprietary FrameCell trunking. Common channel signaling (CCS, Bellcore's term for Signaling System 7, SS7) protocols are supported directly on the edge switch; per-call routing decisions are based on dial digits. End-to-end messages associated with an existing connection are passed transparently from PBX to PBX. Routing completion or failure feedback (individual channel "busy-back") are provided to the originating PBX for individual calls to allow alternate rerouting to be performed by the originating PBX.

In contrast, in traditional PBX networks operating over multiplexer networks, calls that tandem through PBXs will potentially consume more network bandwidth than those that do not. In extreme cases, the network design could allow many times the original call bandwidth to be consumed by a multihop call.

There are other benefits associated with the removal of the tandem function from the voice network. It simplifies the routing tables of the access PBX nodes as dynamic SVCs simplify PBX call-route specification. There is only one connection into the network "cloud." Any call not recognized as "local" is sent to the wide area backbone for routing. There are also reduced requirements for equipment and facilities as the number of interfaces required is dictated by total bandwidth rather than by multiple trunk groups to remote PBXs.

Another part of the voice-processing package evolves the recognition—and technical accommodation—that much of the current traffic on voice networks is not voice calls. Nortel's sophisticated voice adaptation techniques include features for echo cancellation, fax and modem support, and speech activity detection.

PBX features and their successful translation are another part of the package. Nortel's Meridian UIPE (universal ISDN protocol engine) provides transparent support of advanced PBX features across the network via multivendor interoperability. Also, the Passport's ability to intelligently interpret PBX signaling, in conjunction with its inherent dynamic routing capabilities, lays the groundwork for advanced networking services (e.g., negotiation of QoS and call parameters) and path-oriented routing and congestion management.

Finally, there are several out of sight, out of mind (but essential) technical features that support Nortel's ATM approach to voice support. One involves timing accommodation. Nortel's edge switch, the Passport, is synchronized to a common reference clock source. Should the clock fail or prove to be of poor quality, the Passport is able to synchronize the links to its own high-quality internal clock to minimize timing variations. Another is m-law/A-law conversion. u-law is the PCM voice coding and companding* standard used in Japan and North America for T1 ($8,000 \times [8 \times 24 + 1] = 1.544$ Mbps); A-law is the European standard for E1 ($8,000 \times 8 \times 32 = 2.048$ Mbps). Conversion between the two digital channel coding schemes per G.711 is a telephony fact of life.

From a voice efficiency point of view, there is a continuum. Least efficient is using ATM's CBR where an uncompressed voice call takes up about 74 Kbps—64 Kbps for the voice signal and about 10 Kbps for network overhead. Then comes "toll quality" PCM voice at 64 Kbps. Next and much used in TDM "smart mux" networks is 32-Kbps ADPCM, which was designed specifically to operate in multihop PBX environments. (The use of more exotic compression protocols, involving large processing delays, has resulted in poor or inconsistent end-to-end voice quality when employed in anything more than point-to-point voice environments. As a consequence, the ability to limit the number of voice compressions and decompressions to one has become a prime driver in voice switching.) Nortel's "virtual tandeming" enables a more aggressive approach toward compression, resulting in maximized compression ratios of up to 8:1. In the future, there will likely be lower, usage-based ATM tariffs for SVCs; currently, 256-Kbps PVCs are employed. Additional cost savings are likely with usage-based SVCs where multiple network access charges can be avoided.

9.3 MAGELLAN MULTIPLE PRIORITY SYSTEM (MPS)

The Magellan Multiple Priority System (MPS) is Nortel's operating system for its Passport, Vector, and Concorde ATM switches. Its business objectives are suitably lofty: to allow service providers to "optimize network efficiency, allocate resources flexibly, create customized service offerings, and devise variable cost structures for users" [5]. Incorporating prestandard dynamic routing from FORE, it segregates connections for

---

* Companding is the process in which the dynamic range of a signal is reduced for transmission purposes and then expanded to its original value for the end user.

data, video, image, and voice traffic inside the ATM switch so that various applications with different transmission requirements can be handled according to their priorities despite systemic congestion.

Nortel offers a number of useful distinctions in comprehending the breadth of traffic management systems. They propose a three-level hierarchy. At the highest level, there must be node-level controls that operate in real time (in cell times). Next, there must be network-level controls that operate in near-real time (in propagation times across the network, called duration times). In particular, network-level congestion thresholds with hysteresis (state of being behind or late, the lag in response exhibited by a body in reacting to changes in the forces) must be programmable for every queue and the global memory pool. Finally, there must be network engineering tools that support service and network management operating in non-real time.

MPS's call admission control (CAC) converts QoS requests signaled across the UNI into the following VC parameters:

- *Service category:* Constant (CBR), variable (VBR), available bit rate (ABR), and unspecified (UBR);
- *Traffic descriptors:* Peak, sustained, minimum cell rates (PCR, SCR, MCR), cell delay variation tolerance (CDVT), and maximum burst size (MBS);
- *QoS objectives*: Cell-loss ratio (CLR), cell transit delay (CTD), and cell delay variation (CDV).

Further, the Magellan switches support six or more emission and discard priorities. Emission priorities indicate the urgency of a cell; discard priorities indicate the importance of a cell. This allows great flexibility in creating—and supporting—standardized classes of service.

MPS' network engineering tools are extensive and include the all-important billing support. The MPS network management features include the following:

- A graphical user interface (GUI) running Motif (color display, active alarms display, network topology map) and providing operator access security;
- Service management supporting customer-oriented virtual private networks (VPNs) and connection provisioning;
- Configuration management with autodiscovery and provisioning of equipment and preprovisioning options;
- Performance management supporting usage counts and performance monitoring with alerts with thresholds;

- Fault management with network and nodal fault displays and history logs;
- Billing with EBAF (extended Bellcore AMA [automatic message accounting] format);
- E.164 and network service access point (NSAP) address assignment;
- Link capacity management;
- Operator override of automatic routing.

9.4 NORTEL'S CONCORDE ATM SWITCH

The Concorde is the top of Nortel's Magellan line of ATM switches. It is optimized for high-speed cell transport across fiber-optic facilities; its mission is to provide the ATM backbone for the broadband multimedia networks of tomorrow. Figure 9.1 indicates its intended place in Magellan networks.

By the standards of large switch systems of even a few years ago, the capacity is huge: it is rated at 40 Gbps and is scaleable down to 10 and up to 80 Gbps. The basic 10-Gbps Concord (pictured in Figure 9.2) provides up to 64 OC-3 ports or 16 OC-12 ports, which can be provisioned fully duplicated or simplex. Each OC-12 port handles one STS-12c (622 Mbps) path or four STS-3c (155 Mbps) paths. Each OC-3c port handles one STS-3c path. Support for DS1 and E1 interfaces and common channel and channel associated signaling (CCS and CAS) is also provided.

A summary of the Concorde's interfaces/services support appears below:

- DS-3, E3, OC-3c, STM1;
- OC-12, STM4 ATM support;
- VBR/CBR/UBR/ABR;
- PVC/SVC;
- Cell relay service (CRS);
- ATM UNIs: DS-3, E3, OC-3c, STM1, OC-12 STM4;
- ATM NNIs: OC-3c, STM1, STM4;
- Point-to-point and point-to-multipoint;
- Trunking:
    OC-3c/STM1 NNI (1 + 1 optional protection)
    OC-12/STM4 NNI (1 + 1 protection)
- Capacity: 10 to 80 Gbps scaleable.

Not surprisingly, Concorde uses a distributed multiprocessor hardware architecture. Switching elements (SEs) perform ATM switching and transport while real-time controllers (RTCs) do the management of

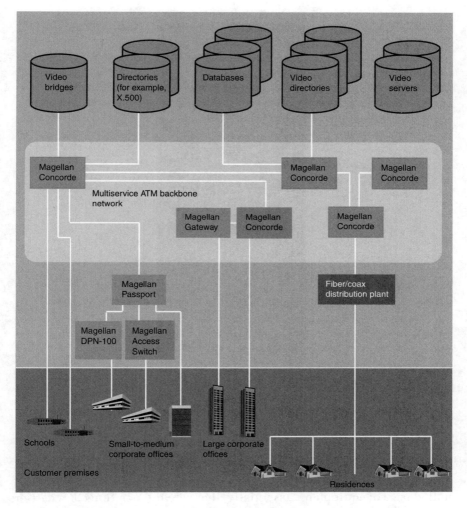

Figure 9.1 Magellan switch hierarchy. (*Source:* Bourne, H., and I. Cunningham, "Implementing the Information Highway," *Telesis*, No. 98, May 1994, p. 11.)

signaling, routing, and connectivity. The Concorde System Manager (CSM) provides for the provisioning of user interfaces and operations, administration, and maintenance of the Concorde system.

The Concorde's software side is consistent with the Magellan philosophy (moving software-intensive services to the periphery) but also buttresses Nortel's claims to "carrier-grade reliability." By limiting the Concorde to ATM transport and not supporting a large diversity of non-ATM services, Nortel minimize the total software volume and rate of

Figure 9.2 Magellan Concorde. (*Source:* Jeanes, D., and M. Unsoy, "Enterprise Solutions for Carrier Networks," *Telesis,* Issue 100, Oct. 1995, p. 23.)

software change; this, in turn, enhances the reliability of these switches and all services supported by the backbone.

Concorde's traffic and network management capabilities are designed to fit in a large-scale carrier environment with a considerable installed base of equipment. Essentially, there are three options. For carriers who already have a considerable Nortel component, there is NT's Magellan Network Management System (NMS), a vendor proprietary system spanning all functional areas of OSI framework for network management, which provides integration between NT's DPN-100 data switch and the Magellan line. Magellan NMS runs on SunSparc workstation 5 and 20 with Solaris and the OSF Motif toolset. For the growing constituency of SNMP managed systems, BNR has developed a SNMP-based interface that incorporates many of the NMS capabilities. A final option is adaptation with a third-party platform already in place.

Management services with the Concorde span all of the "ilities" one expects with carrier-oriented gear. They include the following:

- SNMP and CMIP support;
- Remote access;
- Autodiscovery and provisioning;
- Preprovisioning;
- Per-connection policing and accounting;
- Performance management;
- Billing management;
- Fault management.

Another area where Nortel has expended considerable ingenuity is in the switch packaging for the Concord. Although a function of the compactness of single-mode fiber and the fewer number of physical interfaces needed, it is ironic that physically large, multigigabit ATM switches are tiny when compared to a 6,000 line central office switch where the preponderance of bandwidth is in 64-Kbps units.

Concorde uses a 7-foot standard rack, and multiples of them, featuring front-access 300-mm by 600-mm bays compliant with European Telecommunications Standards Institute (ETSI) physical packaging requirements. There is front access to all cabling connections and field-replaceable components such as processor cards, power supplies, the air filter tray, and the cooling unit. All of these components are connectorized modules that can be removed from the shelf without having to unbolt them or undo cabling hookups. Latches on the cards and flip-down handles on the power packs and cooling unit make for easy removal, and replacement modules are simply plugged back into the shelf. LEDs glowing green, yellow, or red provide status indicators on the face of each cabinet and on the components (cards, power supplies, cooling units) inside.

Also part of the Concorde is an innovative fiber-optimized packaging design. Along with the proverbial fiber-seeking backhoe, Nortel has recognized that many in-service faults have been caused by unintentional mishandling of the optical fiber during card replacement. Concorde's "hands free" fiber-handling systems maintain the minimum bending radius and eliminate the use of plastic tie wraps, often a source of fiber breakage due to overtightening. Nortel predicts that their fiber-optimized packaging design, when compared with traditional fiber-connection techniques, will save carriers up to 30% of the lifecycle cost of maintaining equipment.

9.5 FUJITSU CORPORATION

Headquartered in Raleigh, North Carolina, Fujitsu Network Communications was established in 1988 as a wholly owned subsidiary of Fujitsu

Limited. The parent company is a major player in the global market for computer, semiconductor, and telecommunications products with $36 billion in sales in 1995. For its part, Fujitsu Network Communications develops and markets its FETEX-150 switches for public networks as well as ISDN terminals, terminal adaptors, and broadband multiplexers.

9.6 FUJITSU'S FETEX-150 ESP ATM SWITCH

Fujitsu's FETEX-150 ESP is currently the largest capacity ATM switch on the market, scaleable in 2.5-Gbps units to 160 Gbps. Launched initially in 1989, the latest model, the FETEX-150 ESP, reached full production in early 1996. Now in its self-proclaimed "third generation," the "ESP" stands for "enhanced switching platform. The ESP is also is one-third the physical size of previous FETEX-150 switches and requires less energy to run [6]. A console-level view of the FETEX-150 ESP appears as Figure 9.3.

9.7 DESIGN OBJECTIVES

The target customers for the FETEX-150 are the world's telephone and cable TV distribution companies. Accordingly, the design of the FETEX-150 reflects certain objectives:

- *Multiple application roles*. The FETEX-150 is capable of supporting huge volumes of voice, data, or video traffic as a network "core" switch. In special-purpose, "overlay" networks, it can be deployed in much smaller configurations to serve in frame relay, cell relay (ATM), or video roles, or a mixture of the three. Accordingly, it supports VPC, VCC, PVC, and SVC connection types and UNI 3.1, B-NNI, and B-ICI 1.1 for interoperability.
- *Scalability*. The FETEX-150 can be scaled from a single-stage, single-section supporting traffic to 2.5 Gbps to a three-stage, eight-section multistage self-routing (MSSR) configuration supporting traffic to 160 Gbps.
- *Cost/efficiency*. The FETEX-150 offers several value-added features that allow the service provider to optimize his network. One is the optional traffic concentration. The FETEX-150 concentration option allows the service provider to statistically multiplex input traffic in order to take full advantage of the switching matrix. For example, OC-3c signals that are not fully loaded can be concentrated as much as 4:1 in order to maximize matrix efficiency. A second is the elimi-

Figure 9.3 Fujitsu's FETEX-150 ESP ATM switch. (*Source:* Fujitsu Network Communications, "FETEX 150 ESP ATM Switching System," Aug. 1995.)

nation of SONET multiplexer equipment and self-optimizing buffers. When SONET OC-3c and OC-12c ports are deployed, the carrier has the option of carrying traffic directly out of the office, eliminating the need for external multiplexing equipment. The OC-3c supports medium- and long-reach distances while the OC-12c interface supports short- and long-reach distances. For certain locations, this feature can provide considerable cost savings. A third is self-optimizing buffers, which dynamically allocate (and conserve) switch buffer resources.

- *Reliability.* In the switch proper, the ATM switching matrix, the main processor and call processor are all duplicated. The processor equipment shelf is redundantly equipped and contains a Hyper-SPARC processor. On the interface side, high-capacity SONET interfaces (OC-3c and OC-12c) can be provisioned for 1 + 1 protection.
- *Accounting.* The FETEX 150 performs record data collection and interfaces to a Bellcore-compliant automatic message accounting teleprocessing system (AMATS) for usage-based billing.

- *International switch management standards.* The FETEX-150 conforms to information networking architecture (INA) standards. The system is administered with a management information base (MIB) employing managed objects (MOs) using the common management information protocol (CMIP) interface; it provides the functionality of a common management information service element (CMISE) in transporting network management information. The FETEX also interfaces to public network support systems, including local and remote craft terminals and supports switching control center system (SCCS)-emergency action interface channel (EAI), SCCS-critical indicator channel (CI), and SCCS-maintenance channel (MTCE) capabilities using a standard human machine language (HML) interface; operations systems/network element (OS/NE) CMIP for network provisioning; and TL-1 (transaction language 1, a subset of the ITU-T's man-machine language), a machine-to-machine communications language [7].

9.8 TECHNICAL DESCRIPTION

Schematically, the FETEX-150 consists of three subsystems: the control and processing subsystem, the signal path subsystem, and the maintenance and operating subsystem [8]. The overall switch organization appears in Figure 9.4.

On the FETEX-150, incoming traffic is multiplexed on one of sixteen 622-Mbps "highways" into the switching matrix (ASSW) via a broadband remote line concentrator (BRLC).

The ASSW is a fully redundant, 10-Gbps self-routing 4 × 4 ATM matrix, with each input to the matrix running at 2.5 Gbps. (This internal network highway is designed to switch a 2.5-Gbps payload, making the expansion to higher speeds, such as OC-48 the next logical step.) Each input to the switching fabric contains a multiplexer so that the total 10 Gbps is accepted as 16 network ports, each operating at 622 Mbps.

The BRLC consists of subscriber interface equipment (DS-1, DS-3, OC-3c, OC-12c) and the multiplexing equipment necessary to concentrate and deliver the offered traffic to/from the switch. If the total subscriber bandwidth exceeds 622 Mbps, then the multiplexing equipment provides access concentration.

Each of the four classes of ATM services—CBR, VBR, UBR and ABR—is handled differently internally. Each is assigned a different size buffer and the switch automatically adapts the buffer space to the type of traffic.

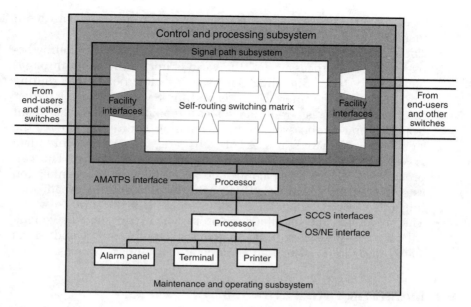

Figure 9.4 Fujitsu FETEX-150 ESP ATM switch schematic. (*Source:* Fujitsu Network Communications, "FETEX 150 ESP ATM Switching System," Aug. 1995.)

The BRLC and the FETEX-150 switch are connected via one or more "umbilical" facilities where the total bandwidth of all the umbilicals cannot exceed 622 Mbps. Along with the service traffic, information is transported on the umbilicals to support the operation, administration, maintenance, and provisioning (OAM&P) features for the BRLC interfaces (subscriber information, alarms, usage measurements, performance monitoring).

Four kinds of processors are employed on the FETEX-150:

- The main processor has a single central processor with duplicated equipment. Its primary function is to coordinate the processing functions and assist in maintenance and operations functions. It also supports the automatic message accounting teleprocessing system (AMATS) function. If the processing requirement exceeds the capability of an uniprocessor configuration, the central processing program functions are split into a main processing function and a call processing function.
- The call processor has a single central processor with duplicated equipment for high availability and performs common call processing. It consists of central processor units, interface cards to connect

to the main processor and the signaling controller, and both main memory and hard disk units.

- The signaling controller provides ATM conversion for signaling between the call processor and the various microprocessors throughout the system. Signaling is used to download the routing information to the interface cards and the SMDS message handler. Information such as performance measurements and alarms are sent from the microprocessors to the signaling controller. There, the signal is terminated and ATM cells are converted before data is sent to the call processor for processing and storage. The call processor and signaling controller also manages the connection establishment/teardown for SVCs using Q.2931 and the broadband-integrated services digital network user part (B-ISUP) protocols.
- The operations and maintenance processor provides extensive logical/physical fault reporting, displays of configuration and alarm status, and interfaces to craft and operations support systems.

9.9 MARKETING THE FETEX-150

One of the most interesting questions relative to Fujitsu's FETEX-150 relates to marketing: given the relatively infant state of ATM today, how do you sell a 160-Gbps switch? In the earlier discussion in this chapter on the Nortel Concorde, it was clear that Nortel was not waiting for the video revolution to fill up those big-backbone bandwidth bundles; a significant part of the Nortel strategy was directed toward the more efficient handling of voice, practically everyone's largest single legacy application. The Fujitsu strategy has been different, and potentially riskier. Although Fujitsu has teamed with partners (Bay Networks) and a major BOC (NYNEX) to employ low-end FETEX-150s for frame relay and/or Internet/ATM transport, by far its current emphasis has been on developing video applications.

The following sections will overview the Fujitsu marketing strategy relative to video, frame relay, and Internet transport.

9.10 VIDEO SERVER/THE NORTH CAROLINA INFORMATION HIGHWAY (NCIH)

Fujitsu has been a major player in one of the most ambitious ATM-based infrastructure projects to date, the North Carolina Information Highway (NCIH). In conjunction with BellSouth, GTE, and Sprint/Mid-Atlantic, Fujitsu has installed twelve of its FETEX-150 ATM switches and seven SMX-6000 service multiplexers, and the consortium has created dozens of subscriber access points. The NCIH currently provides circuit emulation service (CES), SMDS, and FRS via SONET OC-3c [9].

A combination of state funding and donated services and equipment, the NCIH has been touted as a "go-ahead" infrastructure project. Technology publicists hail "ATM...the high-speed "carry all technology" and the state of North Carolina...as the all-important "anchor tenant" [10]. Governor Jim Hunt of NC modestly notes that "This first Information Highway gives North Carolina tremendous opportunities to serve all of our people across the state" [11]. Also noted is that North Carolina was listed as the number one state for new and expanded corporate facilities for 1992 by *Site Selection* magazine.

In fact, the NCIH did not spring from the private sector, much less the State of North Carolina. Its roots can be found in Vistanet, a 5-Gbps test bed funded by the Department of Defense's Advanced Research Projects Agency and the National Science Foundation in 1989. Vistanet brought together BellSouth, MCNC, GTE, Fujitsu, and the University of North Carolina-Chapel Hill. The initial application was telemedicine—specifically, advanced radiology treatment—for military bases and rural hospitals. Another pre-NCIH project was "Vision Carolina" in 1991. In a three-year trial, BellSouth linked 16 educational sites on a fiber-optic network. One hundred distance learning classes per week were conducted over a DS-3 (44.7 Mbps) infrastructure.

By 1992, a consortium comprising the State of North Carolina, telcos, and equipment vendors were confident enough to launch the NCIH. Nonetheless, costs have remained high. With the State funding the initial costs of NCIH access, first sites cost $105,000 each to connect. As more sites have come online the costs have fallen to about $80,000 per site. In addition to a one-time connection fee, the State charges school sites approximately $4,000 a month for OC-3c service, a stratospheric amount in the educational community [12].

Intriguingly, the focus has recently shifted to an application area more politically au courant—criminal justice. In an application note, Fujitsu observes that "North Carolina currently spends more than $5,000 per prisoner for an arraignment in which a judge reads the charges and the prisoner enters a plea... the state could save as much as $4 million annually through remote arraignments" [13]. Video arraignments are catching on; today, they are being used in Ohio, Texas, Arkansas, California, North Carolina, and Washington, DC Via video, an arraignment takes minutes versus a system where the prisoner has to be transported 30 miles and then wait his or her turn for arraignment. There are also the intangible, but politically important, public safety factors in reducing the transport of prisoners. Other penal applications beckon, such as telemedicine, enabling remote diagnostics for prison inmates. The application note observes that "few physicians choose to live near correctional institutions ...[and] physicians travel and other related expenses...[as well as] transporting inmates to a remote medical facility is very expensive..." Further, "at least two guards are required for security, plus a state vehicle, in addition to the medical fees...[bringing] the average cost nationally [to] $700 per prisoner trip to a medical facility." Another potential criminal justice application is education, both for training in criminal justice, where the state could save nearly $1 million annually by offering remote site training to the more than 10,000 students that attend the academy each year, and in offering high school and college-level courses to prisoners.

Ironically, outside the criminal justice system, distance learning—the granddaddy of "closed circuit" video applications, and the one with the longest track record—does not appear likely to launch a thousand ATM switch ships. A recent study by DRI/McGraw-Hill examined the cost savings associated with the adoption of telecommunications technology such as broadband ATM and found a potential national cost savings of only 4.5% annually [14].

Where does the video market leave Fujitsu and other large ATM switch makers? As long as the United States trails only the former Soviet Union in numbers incarcerated per capita, it would seem that criminal justice applications employing video (arraignments, medical exams, education) would be sure winners. With the cost and organizational realignments underway in the health care industry, telemedicine would also seem likely to carve out an increasing niche. High bandwidth prices will probably continue to hobble WAN videoconferencing and less-focused distance learning. Although an economic development application—whereby a client takes a live, directable video tour of a potential site—has been hyped, the advantages of "being there" seem overwhelming. To this observer, the sleeping giants of video applications lie in entertainment—specifically, video-on-demand (VoD) and some kinds of interactive video, most likely remote shopping and game playing. All have well-established financial track records and are only limited by the availability of suitably alluring "software" (films, sports, games, merchandise).

9.11 FRAME RELAY/NYNEX

A part of Fujitsu's marketing strategy has been to sell the FETEX-150 to service providers as a frame relay switch. In 1993, NYNEX, since merged with Bell Atlantic, selected FNS as its first commercial broadband ATM switch vendor. Upon completion of trials, NYNEX expects to commercially deploy the FETEX-150 to expand its frame relay capabilities as well as introduce additional fast packet services, such as ATM cell relay service and SMDS.

Unlike the video market, the frame relay market is dominated by the (relatively) impecunious. Andrew Greenfield of StrataCom, now part of Cisco, has observed that "narrowband services like frame relay [are] what most people can afford and that's the amount of bandwidth that most people need for their current applications..." Furthermore, "up to a certain point...roughly the DS-1 rates, access for data is maybe better using frame relay...[where] above DS-1, it's advantageous to convert frame relay to ATM..." [15]. Less clear is whether the FETEX-150 is appropriate for the highly competitive FR market and whether Fujitsu can become a significant player without more carrier partnerships.

9.12 CELL RELAY/BAY NETWORKS

In a move paralleling the NYNEX alliance, FNS in 1995 paired with Bay Networks to provide an Internet access "package" for Internet service providers (ISPs). Bay Networks, the product of a merger of Wellfleet and SynOptics, is an aggressive number two behind Cisco in the router market. As the ISP package is envisioned, Bay Networks BNX routers would consolidate the various input traffic types and funnel them over an ATM UNI to the Fujitsu switch. In this union, the Bay Networks equipment provides the multipurpose access for users to connect to the Internet, including dial-up modems, ISDN basic rate, leased line at a variety of rates, and packet (frame relay, X.25). With Internet traffic consolidated, the FETEX-150 provides the authentication/accounting service of the ISP and the FR or ATM connections to the Internet [16].

Although too recent to evaluate its success, providing a high-capacity, easily expandable "back end" for ISPs could be a nice niche for the FETEX-150. Few ATM switches on the market today have serious accounting capabilities and that, combined with the FETEX-150's ability to grow in small increments to a very large size, makes it a natural for this fasting-moving application area.

9.13 SUMMARY

This chapter has overviewed two vendors offering very large ATM switching platforms; the focus has been not only on the technologies involved but where they see the opportunity in broadband applications. Other than selling to carriers, their respective marketing plans are very dissimilar. Nortel, with its Concord switch, hopes to leverage its traditional strengths—voice processing expertise, long experience in the carrier environment, and worldwide distribution channels and service. Fujitsu, with its FETEX-150, has vigorously pushed the expansion of new video applications. It has worked from alliances such as the NCIH; the degree of its success is likely tied to the rate of the expected cost declines in broadband WAN bandwidth. More modestly, it has also aimed at popular niches, such as carrier-provided FR/SMDS/ATM services and provisioning ISPs.

The differences have not been limited to marketing. In technical approaches, the roles have been reversed. Nortel has been aggressive in offering technically elegant "prestandard" software and pushing "value-added" technologies. Fujitsu has played it conservatively and stressed its support of, and interoperability with, a wide variety of carrier standards. Nortel plainly hopes its innovations, particularly in voice processing, will sell more switches and offset the added costs of maintaining proprietary and standards-based software versions.

End Notes

[1] Csenger, M., "Northern Covets Enterprise ATM," *Communications Week*, May 2, 1994, p. 143.

[2] Pappalardo, D., "ATM Products Ease Migration," *Communications Week*, March 27, 1995, p. 8.

[3] Greene, T., and D. Rhode, "Sprint Pushes Nortel ATM Switch to Edge," *Network World*, Nov. 13, 1995, pp. 21, 28.

[4] Taylor, K., "The Ultimate ATM Mix: More Talk, Less Bandwidth," *Data Communications*, Jan. 1996.

[5] Nortel Magellan, "Magellan MPS: Nortel's Traffic Management System," www.nortel.com/

[6] Fujitsu Network Switching, "Fujitsu Network Switching Announces New ATM Switch," Aug. 31, 1995.

[7] Fujitsu Ltd., FETEX-150 ESP, Public Network Broadband Switching System, Description and Analysis, Raleigh, NC: Fujitsu Ltd., 1995, p. 6.

[8] Ibid., pp. 12–18; Babson, B. D., G. DeVal, and J. S. Xavier, "ATM Switching and CPE Adaptation in the North Carolina Information Highway," *IEEE Network*, (Vol. 8, No. 6, Nov./Dec. 1994, pp. 40–45.

[9] Williams, T., "Hitching a Ride on the NCIH," Special to *Telephony*, April 24, 1995.

[10] Egolf, K., "Riding in the Express Lane,' Special to *Telephony*, April 24, 1995.

[11] Fujitsu Network Switching of America, "What's Going On Here?," Fujitsu Ltd., 1993, unpaged.

[12] Williams, T., "Hitching a Ride on the NCIH," unpaged.

[13] Fujitsu, "Broadband ATM Applications for State Networks," http://www.fns.com/announce/appnotes/note3.html, July 11, 1996, p.4.

[14] Ibid., p. 5.

[15] Watson, S., "Have ATM, Will Travel," Special to *Telephony*, April 24, 1995, unpaged.

[16]   FETEX-150 ESP Switching System, Internet Access and Traffic Management and FNS, http://www.fns.com/announce/appnotes/note2.html, July 11, 1996;
FNS, "Fujitsu and Bay Networks Announce Joint Development and Marketing Alliance," http://www.fns.com/announce/pr/9510301.html, Oct. 30, 1995.

The ATM Prognosis **10**

In an industry that moves so fast that, hyperbolically, "if you go for lunch, you become lunch," prognosis is a risky business. Yet ATM as an intellectual conception dates from the late 1980s and commercial ATM products have been on the market since 1992. Much has happened and much remains to happen. Although the connective thread may appear somewhat strained, in this last chapter several subjects will be discussed: the ATM Forum's recent "Anchorage Accord" on new standards, the rise of fast Ethernet and the promise of gigabit Ethernet, the standards process versus the market, and ATM in the longer term.

10.1 THE "ANCHORAGE ACCORD"

At the ATM Forum's mid-April 1996 meeting, the group agreed, in what has been called the "Anchorage Accord" or "Anchorage Freeze," to a major slowdown in the ATM standards process. Observers have noted that since 1991, the Forum has approved nearly 50 specifications, that 20 more are due by the end of the summer of 1996, and dozens more are in progress. This pace has proven hard on the vendors and hell on the buyers.

The Accord's slowdown involves several sensible courses. First, backward compatibility will be required for all future releases. (UNI 3.1, for example, was not backwardly compatible with UNI 3.0.) Second, it identifies two classes of specifications—"foundation" and "expanded features." Foundation specifications, which will be not fully defined until 1997, will include the core LAN and WAN ATM specifications, such as UNI signaling, the private network-node interface (P-NNI), the broadband intercarrier interface (B-ICI), traffic management, and key physical layer interfaces. Expanded features will include areas that need addressing—

or modification—as soon as possible, such as LAN emulation, circuit emulation, audiovisual multimedia services, and voice over ATM.

However well meant, the Accord does not solve many of the salient problems. As Mary Petrosky of The Burton Group, one of the most sagacious of ATM watchers, has observed, the process is inflicted with hysteresis, the state of being behind or late [1]. Already noted is that the foundation specification will not be fully defined until early in 1997. Once that occurs, compliant products will not be available for another 6 to 12 months. The more complex standards (UNI, PNNI) will probably take two years to go from specification to the fielding of fully compliant versions, although PNNI subsets are available today. Further, many of the ATM specification are interdependent, and a change in one requires a change in others. Additional hardware upgrades are also in the offing. Many current switches will require additional memory and processor capacity to support the new rate-based traffic management scheme [2]. Finally, nothing guarantees that the vendors, who have their own priorities as well as proprietary solutions at stake, will be expedient in supporting the new standards.

10.2 "FAST" ETHERNET AND THE PROMISE OF A GIGABIT ETHERNET

In the nexus of fast-moving technology, one of the fastest has been fast Ethernet, a 100-Mbps follow-on of the now-famous Xerox PARC creation of the 1970s. Technically, fast Ethernet dates from June 1995, with the adoption of the IEEE 802.3u (100BaseT) standard. It presumes a star typology and consists of adapter cards, repeaters (often incorporated in the switches), switches, and routers. There are three cabling options: 100BaseTX runs on either 2-pair CAT5 or 2-pair STP; 100BaseT4 runs on 4-pair CAT3 or, with 4B/5B signaling, 4-pair UTP; and 100 BaseFX runs on 62/125-micron multimode fiber. 100BaseTX and 100BaseFX are the most popular. There are a large of number of hybrid cabling/repeater options including, at the high end, full duplex transmission to 2,000m. Other fast Ethernet benefits include increased VLAN support, auto-sensing interfaces, autospeed arbitration, buffering configuration options, level-three interaction, and prioritization.

A strong proponent of fast Ethernet has been the microprocessor giant, Intel. Along with touting their new PCI bus which runs at 1.56 Gbps, Intel points out that their new, two-speed 10/100 (Mbps) NICs are only "the cost of a pizza" more than predecessor 10 (Mbps) NICs.

Actually, for users who can make do with 100 Mbps or somewhat less—and without strict QoS—there is a sort of Chinese menu of minimally disruptive and less expensive (than ATM) performance upgrades for today's Ethernet LAN: shared Ethernet to microsegmented shared Ethernet to switched Ethernet to shared fast Ethernet to switched fast Ethernet.

It is clear that fast Ethernet is clearly outselling ATM today in the LAN market. It provides a number of (see above) low-cost and minimally disruptive options for today's legacy Ethernets to increase performance. Particularly salient is the low cost of 10/100 NICs, already under $150. It is not, of course, as cheap as it appears. As the Yankee Group has computed, for the "WinTel" software/hardware environment, it costs $500 per workstation ("seat") to swap out a network interface card (NIC), so selecting NICs (10/100-Mbps Ethernet versus 16-Mbps token ring versus 100-Mbps F/CDDI versus 25/100/155-Mbps ATM) on small cost differences in NICs is probably myopic.

The gigabit Ethernet technology, represented by the Gigabit Ethernet Alliance, hopes to have a spec by late 1998, and a parallel effort is underway by the IEEE on Gigabit VG-AnyLAN [3]. Gigabit Ethernet appears to have a good chance of beating out ATM in the LAN market, at least in the early going, and for many of the same reasons as 100-Mbps Ethernet: Ethernet is a mature, simple, well-understood, and cheap technology; there is no consensus that ATM is any of the above.

A more interesting question concerns the degree with which ATM's more sophisticated capabilities can be duplicated by evolutionary LAN technologies. For many future applications contemplated today, simply going faster at low cost is not enough. A number of related efforts are going beyond the higher speeds promised by fast or gigabit Ethernet. Switched Ethernet is offering dedicated bandwidth. Isochronous Ethernet, IEEE 802.9, and 3Com's PACE technology are all promising isochronous support. Virtual LANs tout their increased ease of management. And enhanced protocols such as Ipng and RSVP promise multicast and bandwidth reservation [4].

On the Ethernet front, professional LAN watchers are particularly bullish on fast Ethernet's technical advantages over its 100-Mbps competitor, 100VG-AnyLAN. The Erin news article cites Dataquest on the Ethernet installed base as being 70M, and that fast Ethernet (100Base-T Ethernet) will reach 4.2M, and 100VG-AnyLAN 780K, in 1996; by 1997, the predictions are 10.6M for fast Ethernet, 1.7M for 100VG-AnyLAN.

10.3 STANDARDS, THE FORUM, AND THE MARKETPLACE

The development of ATM has set a number of thoughtful observers to thinking about process—how might a promising technology be best developed and fielded? There are, of course, at least three ATMs. There is the ATM contemplated by the original standards writers in 1988. There is the ATM developed by, and still in process, by the ATM Forum since 1991. And then there is the ATM in the products, first appearing with FORE in 1992.

Nortel characterized the first group's perspective when they wrote that "ATM combines all of our historical knowledge about representing and moving information in a single technology"[5]. The Forum, whose inner group is dominated by makers of routers, hubs, frame relay switches, TDM multiplexers, and so forth specified the ATM technology, but also partitioned it such that their products, present and future, would "fit." Then the commercial ATM switch vendors made it useful, but usually with proprietary bells and whistles. FORE ensured that legacy LANs could run over it, and run much faster. Newbridge spawned a new generation of smarter "smart" multiplexers. StrataCom and Cascade used it to extend their frame relay offerings. Nortel married it to their voice expertise to save users substantial dollars in circuit recurring costs.

A number of observers have found this process less than optimal. There have been negative illusions to OSI, and, to a lesser extent, ISDN. A columnist for *Data Communications*, Robin Layland, particularly appalled by the melee over congestion control—which no one is absolutely sure will work to specification—has argued that the Forum should follow IETF practice and require a real product implementation prior to proposing a standard [6]. He further goes on to criticize the ATM Forum for an AAL 3/4 specification that is "unrelated to how computers work," an AAL 1 spec that "wastes bandwidth," and a QoS spec that is "too complex" to implement.

There is considerable merit in his criticisms. Certainly AAL 1 has limited utility except for backward compatibility with TDM multiplexer networks. AAL 2 needs efficient specifications for voice and video. AAL 3/4 is probably dead unless SMDS catches on soon. And AAL 5 is so dominant in the data area that a number of commercial workgroup switches offer nothing else. If ATM is bypassed by other technologies, such as evolutionary LAN technologies, these mistakes, in retrospect, may be perceived as fatal. If not, and new AALs come to take their place, they will be seen as vestigial parts of the great shaggy beast, of interest to few except to those who write technology books.

10.4 THE LONGER TERM

The economist John Maynard Keynes, in savaging the classical economists and their redemptive faith in the long run, was fond of observing that "in the long term, we are all dead." Despite the slippery footing involved with the "short term, long term," a number of observations are germane:

- ATM in the LAN market needs to be cheap and fast. Where it is cheap today (25 Mbps), it is not fast; where it is fast (155 Mbps), it is not cheap. Hovering over the LAN mess is the requirements question: how many LAN users need 155 Mbps and QoS?

    There are two answers here. Today, except for some "customers of the future" like Disney Animation that, for performance reasons, have no alternative, they do not appear numerous, especially at the client side. They are, however, prestigious, and it helps explain Sun Microsystem's recent 622-Mbps NIC announcement. The second answer is that ATM could become popular as the "upstream" technology, where aggregated LAN traffic, lots of it, hits servers and WANs. For instance, most router, hub, and workstation vendors have announced "upstream" ATM interfaces; of the installed base, almost half are 155-Mbps OC-3 fiber [7]. But, at least today, these upstream, server- or WAN-side capabilities do not include an ATM API or deliver ATM QoS.

- ATM in the WAN market will be hard to beat, with FR the only serious competitor. The present alternative technologies—various combinations of TDM multiplexers, routers, FR switches—have lots of negatives:

    > As a "standalone system implements each new protocol scheme separately, it is not aware of the other protocols operating on the same physical network. As a result, each protocol must have its own physically partitioned bandwidth [TDM multiplexer] or compete for available bandwidth across the entire network [routers]...traditional router networks require each node in the backbone to understand all network protocols and maintain forwarding tables for all protocols by continually exchanging control information with the other network nodes" [8].

    Again quoting Nortel, there is what they tout as the "Power of the Cell:" "ATM is the only technology that supports multiple services, handles a variety of traffic speeds, offers both guaranteed bandwidth and statistical on-demand connections, and adheres to stringent delay and delay-variation requirements" [9].

- There will be untidiness for as far as the eye can see. David Stockman, former Budget Director for Ronald Reagan, several years after "cooking the books" with unrealistic projections to claim a balanced budget, later recanted to Congress, and, in a particularly apt phase, confessed to "red ink for as far as the eye can see." With ATM, it is "untidiness for as far as the eye can see."

This untidiness is a symptom of a number of telling points about the relationship between standards, technology and the ATM industry. As Mary Petrosky has observed: (1) even with the Anchorage Accord in 1996, ATM will not be a mature technology until 1999; (2) with its continuing cycle of mergers, buyouts, and licensing deals, the industry is clearly unstable; (3) for the ATM buyer, the expectation should be at least annual upgrades in hardware and software for the foreseeable future; (4) at least into 1997, single-vendor shops will be necessary to ensure that ATM works as advertised; and, (5) in ATM's applications future, whenever immediate functional needs clash with long-developing standards, "prestandard" solutions are guaranteed [10].

End Notes

[1] Petrosky, M., "The Anchorage Accord Promises to Stabilize ATM," *Network World*, May 6, 1996, p. 39.
[2] Tom Ramsey of the MITRE Corporation, a close observer of the dynamics of today's router market, has summarized this phenomenon: One major software upgrade per year, every year. "Hundreds" of bug fixes per year. "Code creep" that doubles the size of the code every two years. Additional memory, both RAM and flash, that must be installed every two years to accommodate the new software versions. And a CPU upgrade every three years to keep up with the new code.
[3] English, E., "Ethernet Looks to a Gigabit Future," *Computer*, June 1996, p. 18.
[4] Cohen, J., "ATM Under the Gun," *Network World*, May 13, 1996, pp. 1, 65.
[5] Nortel, "Implementing the Information Superhighway," www.nortel.com/.
[6] Layland, R., "Can the ATM Forum Avoid OSI-fication?," *Data Communications*, Dec. 1995, pp. 87–88.
[7] Mier, E. R., "ATM At 25 Mbps: What Inquiring Users Want to Know," *53 Bytes*, Vol. 4, No. 1, pp. 4-5.
[8] Nortel Magellan, "Magellan MPS: Nortel's Traffic Management System," www.Nortel.com/.
[9] Nortel, "Implementing the Information Superhighway," www. Nortel.com/.
[10] Petrosky, M., ATM Market Maturing: The Burton Group Report, *Salt Lake City, UT: The Burton Group, v1, May 1996.*

Annotated Bibliography

Abe, S., and T. Soumiya, "A Traffic Control Method for Service Quality Assurance in an ATM Network," *IEEE Journal on Selected Areas in Communications*, Vol. 12, No. 2, Feb. 1994, pp. 322–331. Basically, a two-pronged, simulation-tested, approach to ATM congestion control. Short-term congestion is handled by dumping flagged (low-priority) cells. Long-term congestion is handled by sampling cell traffic and statistically predicting long-term trends from the data, and subsequently practicing "intelligent" call admission.

Abou-Arrage, G., and I. Merritt, "Making the Journey to Enterprise Networking," *Telesis*, No. 98, May 1994, pp. 35–51. A high-level product description of NT's Magellan line consisting of the Concord backbone switch, the gateway, a concentrator and router, and the Passport, an "edge" (of the public network)/"enterprise" (customer premises) switch.

Acampora, A. S., *An Introduction to Broadband Networks, LANs, MANs, ATM, B-ISDN, and Optical Networks for Integrated Multimedia Telecommunications*, New York, NY: Plenum Press, 1994. A medium-difficulty introductory work covering the principles of broadband LANs, packet networks, MANs, ATM, performance issues (admission control, policing, flow and priority control, and self-learning strategies), and lightwave networks.

Adam, J. F., et al., "Media-Intensive Data Communications in a 'Desk-Area' Network," *IEEE Communications Magazine*, Aug. 1994, pp. 60–67. VuNet is an MIT project where several VuNet "desk area" networks are connected by Aurora's SONET/ATM net at 622 Mbps. Contains a number of neat ideas: (1) multimedia devices attached directly to the LAN as ATM devices; (2) software-intensive approach to allow easy accommodation and adaptation of newer workstation and peripherals and "analog"-like graceful scaling and adaptation; (3) VuNet as a high-speed I/O bus for connecting plug 'n play ATM-based peripherals; (4) adding 3 bytes to ATM's 53-byte cells to make a cell "four and eight byte-aligned" vis-à-vis 32- and 64-bit computer buses; (5) video camera of the future outputting an ATM digital video stream; and (6) throughput invisibility LAN/WAN.

Akaike, T., et al., "Distributed Multilink System for Very-High-Speed Data Link Control," *IEEE Journal on Selected Areas in Communications*, Vol. 11, No. 4, May 1993, pp. 540–549. Advocates a protocol control system for very high performance for data communications in a broadband network. It consists of a protocol controller connected in parallel and uses the multilink procedure in a distributed way. A token ring topology and the empty-selection algorithm are used for distributing the transmission allocation function to controllers. On the receiving side, the resequencing function is distributed according to a method in which sequence numbers are broadcast to each controller.

Alexander, P., and K. Carpenter, "ATM Net Management: A Status Report," *Data Communications*, Sept. 1995, pp. 111–116. A good overview by two StrataCom staff on what's available today (the SNMP-based interim local management interface, part of UNI 3.1, for managing interfaces between networks) versus the ATM Forum's planned five-layer management model and OAM facility. The latter defines five key management interfaces, M1–M5. M1 and M2 define the net management system and the ATM switch and are based on the IETF's SNMP-based MIB II and AToM (RFC 1695). M3, the customer network management (CNM) interface defines the interface between customer and carrier management systems. M4 provides network and element management level (NML, EML) views into the carrier's network management system and public ATM network. M5 is the management interface between a carrier's own network management systems.

Alexander, P., "Wanted: New Net Mgm't For ATM," *Communications Week*, June 26, 1995, p. 71. Argues that ATM needs scalable, standards-based, industrial-strength network management, especially in cost allocation and billing, and that SNMP is not up to the challenge.

Alles, A., *ATM Internetworking*, San Jose, CA: Cisco Systems, May 1995. A brilliant and challenging examination of ATM implementation issues by Cisco's ATM product line manager. Strongly router and connectionless network-oriented (he ignores the existence of mux, X.25 or SN A networks!). He makes a number of important points relative to ATM implementation: (1) ATM today will involve the overlay of a highly complex, software-intensive protocol infrastructure; (2) current routable protocols do not map well into ATM or allow ATM's unique QoS properties; (3) argues convincingly that ATM is indeed a "full-fledged network layer protocol—one, indeed, that is perhaps at least as complex as any that exists today;" (4) contains the best extant description of LAN emulation (LANE) ."...LANE is essentially a protocol for bridging virtual LANs across ATM" and runs under ABR/UBR without QoS; (5) says that LANE Phase 1 protocol has scalability problems due to the need for flooding through the BUS and single points of failure; (6) sees the router future tied to multiprotocol over ATM (MPOA) while citing ATM interconnection problems with virtual LAN protocols like LANE, RFC 1577, next hop resolution protocol (NHRP), and MPOA; (7) observes that router costs today are driven by the high-performance processors and memory required for the route processing function, where the packet switching function rides the ASIC cost curve; and (8) contains excellent appendices on ATM traffic management (type of service × traffic parameters × QoS parameters) and ATM standards (ITU-T × ATM Forum × ANSI T1S1 × IETF).

Alles, A., *Next-Generation ATM Switch: From Testbeds to Production Networks*, San Jose, CA: Cisco Systems, 1996. Part a Cisco-Lightstream sales support document, part state-of-ATM in 1996, Alles' white paper is full of interesting observations on the state of the standard and current ATM practice. He (1) argues that most current ATM switches are not ready for production environments, and, because of the new UNI/PNNI/ABR standards, cannot be upgraded; (2) takes a basically data network view of ATM, the ATM world as UBR/AAL 5/LANE without the need—or ability—to do user-initiated QoS; (3) argues the superiority of shared memory switch designs because of their ability to upgrade software; (4) argues the practical need for UNI 3.0/3.1 dual stacks for interoperability with ILMI and PNNI playing major roles in autoconfiguration; (5) stresses the need for high-performance hardware to support signaling, route, and RM processing; (6) provides, on ABR, an excellent discussion of congestion control alternatives—EFCI, relative rate marking (RRM), and explicit rate marking (ERM); and (7) until ABR is available in 1997, claims that UBR with EPD and tail packet discard (TPD) constitutes "UBR+."Anagnostou, M. E., et al., "Economic Evaluation of a Mature ATM Network," *IEEE Journal on Selected Areas in Communications*, Vol. 10, No. 9, Dec. 1992, pp. 1503–1509. Another economic study of B-ISDN, this of ATM economic viability across various business and residential sectors. Questions having to do with investment capital, cash flow, marketing, products, etc. are not addressed.

Anderson, J., et al., "Fast Restoration of ATM Networks," *IEEE Journal on Selected Areas in Communications*, Vol. 12, No. 1, Jan. 1994, pp. 128–138. Points out that the ATM technology offers superior restoration features (ATM cell-level error detection, inherent rate adaptation, non-hierarchical multiplex) over STM nets, but ignores the fact that the one travels over the other.

Aoyama, T., I. Tokizawa, and K. Sato, "ATM VP-Based Broadband Networks for Multimedia Services," *IEEE Communications Magazine*, April 1993, pp. 30–39. The authors, from NIT, look forward to providing "cost-effective, high-speed multimedia lease-line services" using BISDN/ATM technology by the mid-1990s. They plan using a virtual path (VP) concept to achieve this and describe a three-layered architecture of a transparent network of physical media, path and circuit.

Aoyama, T., I. Tokizawa, and K. Sato, "Introduction Strategy and Technologies for ATM VP-Based Broadband Networks," *IEEE Journal on Selected Areas in Communications*, Vol. 10, No. 9, Dec. 1992, pp. 1434–1447. Describes an ATM-based infrastructure for the "future advanced information society" and how the evolution might occur. They propose a virtual path concept (PVC) that would supersede traditional hierarchical synchronous transfer mode (STM) digital cross-connect systems, allowing lower cost and greater service flexibility.

Armbrüster, H., and K. Wimmer, "Broadband Multimedia Applications Using ATM Networks: High-Performance Computing, High-Capacity Storage, and High-Speed Communication," *IEEE Journal on Selected Areas in Communications*, Vol. 10, No. 9, Dec. 1992, pp. 1382–1396. Two telecommunications engineers from Siemens A. G. overview the broadband future and its telecommunications requirements. Conspicuously absent are the applications that are going to drive this infrastructure and pay for it.

Armbrüster, H., "The Flexibility of ATM: Supporting Future Multimedia and Mobile Communications," *IEEE Communications Magazine*, Feb. 1995, pp. 76–84. Written by the vice-chair of the ATM Forum's European Education Committee, the article is one protracted commercial for ATM, as well as long on planning, pilots, and cooperation, and short on attractive applications that people are willing to pay for.

Armitage, G. J., and K. M. Adams, "How Inefficient is IP Over ATM Anyway?," *IEEE Network*, Jan./Feb. 1995, pp. 18–26. Authors put IETF 1483 for IP over ATM to test. Develop a metric, equivalent ATM cells (EAC), and find that efficiency is pretty decent on a large traffic sample (70–80% range), especially with AAL 5 (as opposed to the higher overhead but muxable 3/4); that the 48-byte payload size is pretty good over the mix; and that TCP/IP header compression (40 to 5 bytes!) has a big payoff with small IP payloads.

Asatani, K., "Standardization of Network Technologies and Services," *IEEE Communications Magazine*, July 1994, pp. 86–91. Boring recital of standards efforts on BISDN, IN, landmobile telephone, and audiovisual multimedia. Sees, with unified media, more collaboration among standards groups. Excellent diagram on of ITU-T recommendations on B-ISDN.

Ash, J., and P. von Schau, " Communications Networks of the Future," *International Telecom Report*, Vol. 15, Nov./Dec. 1992, pp. 5–8. Basically a sales job for Siemen's Vision O.N.E strategy—essentially open interfaces and networks capable of offering a broad range of applications. Describes the challenge of evolving from the "plesiochronous" networks of today to the ATM-based networks of the next century. Several good diagrams showing the trends, pieces, and technologies to be involved.

ATM Forum, *ATM User-Interface Specification*, version 3.0, Englewood Cliffs, NJ: PTR Prentice Hall, 1993. Specification describes in great detail how an ATM user accesses (1) a public network or (2) a private ATM switch (and how the private ATM switch accesses the public ATM network). Fred Sammartino's "Foreword" modestly predicts "ubiquitous deployment" for ATM and states that it has "become synonymous with communications in the information age." Also contains a disclaimer that is, inadvertently, a telling indicator of contemporary business values.

ATM Forum, *ATM User-Interface Specification*, version 3.1, Upper Saddle River, NJ: PTR Prentice Hall, 1995. When one realizes this long and complex specification is just a piece, albeit an important piece, of the larger ATM specification in progress, one has conflicting thoughts: how can anything this complex ever work reliably? How can anything this complex ever have a chance of working without such a specification? Contains a number of useful summary diagrams.

Axner, D., "Portrait of an ATM Switch," *Network World*, June 19, 1995, p. 65. Compares four enterprise ATM switches (Cascade 500, GDC Apex-DV2, Newbridge 36170, and StrataCom BPX). Likes the Cascade 500 for its VC capacity (16K per I/O module) and high port density (to OC-12). Also, a good discussion of admission control (CAC) and policing (UPC, NPC).

Axner, D., "What Users Should Know About ATM Carrier Services," *Telecommunications*, March 1996, p. 29–35. Sobering review of what's available from the LEC, IEC, and "competitive access providers" (CAPs) such as LDDS WorldCom (ex-Witel), MFS DataNet, and Teleport. For most, it is PVC-only at T1/T3/OC-3, limited to CBR and VBR, and without SVCs, ABR, LANE, and, charged per month with heavy installation charges, expensive.

Babson, M., et al., "ATM Switching and CPE Adaptation in the North Carolina Information Highway," *IEEE Network*, Vol. 8, No. 6, Nov./Dec. 1994, pp. 40–46. North Carolina's NCIH provides ATM-based service for distance learning in medical and educational application areas in three flavors: circuit emulation, frame relay, and SMDS. The article describes the architecture of the Fujitsu FETEX-150 (and associated equipment) to support these services.

Bae, J. J., and T. Suda, "Survey of Traffic Control Schemes and Protocols in ATM Networks," reprinted in W. Stallings, *Advances in Integrated Services Digital Networks (ISDN) and Broadband ISDN*, Washington, DC: IEEE Computer Society Press, 1992, pp. 219–238. A survey and review article of ATM topics including modeling of traffic sources, congestion control, error control, and priority schemes to support multiple traffic classes. In particular, the authors argue for preventive congestion control and for error control, a block acknowledgment scheme in conjunction with a block-based selective-repeat retransmission protocol executed on an edge-to-edge basis.

Banerjess, S., V.O.K. Li, and C. Wang, "Distributed Database Systems in High-Speed Wide-Area Networks" *IEEE Journal on Selected Areas in Communications*, Vol. 11, No. 4, May 1993, pp. 617–630. Authors identify the issues in developing distributed database systems (DDBS) for applications in a high-speed environment, demonstrate the inadequacy of existing database protocols, and advocate a new concurrency control protocol that performs better than traditional DDBS in a high-speed network environment.

Banks, S., "ATM Virtual Trunking Provides the Best of Private, Public Nets," *Network World*, April 15, 1996, p. 37. Banks, a marketing manager for StrataCom, advocates ATM "virtual trunking," which would allow network managers to build hybrid public/private networks at less cost than leased circuits. The advantages include support of SVCs and the extension of congestion control and other features across the public network. Compares his virtual trunking proposal favorably to UNI and peer-to-peer networking approaches.

Banniza, T. R., et al., "Design and Technology Aspects of VLSI's for ATM Switches," *IEEE Journal on Selected Areas in Communications*, Vol. 9, No. 8, Oct. 1991, pp. 1255–1264. An overview of the design requirements through VLSI of the Alcatel ATM switch, a space division 4 × 4 155-Mbps self-routing, folded fabric, randoming input, fault tolerant "silicon realization."

Banwell, T. C., et al., "Physical Design Issues for Very Large ATM Switching Systems," *IEEE Journal on Selected Areas in Communications*, Vol. 9, No. 8, Oct. 1991, pp. 1227–1238. Authors contend that the switching capacity requirements of future CO switching systems to be 1 Tbps or more. To support this, two to three orders of magnitude increase in throughput and the accompanying increase in complexity, new physical design alternatives are needed to provide rich, high-speed connectivity, excellent thermal management characteristics, and high reliability. In the paper, the authors examine the physical design issues associated with terabit/second switching systems, and especially the customer access portion of the switch.

Barham, P., et al., "Devices on the Desk Area Network," *IEEE Journal on Selected Areas in Communications*, Vol. 13, No. 4, May 1995, pp. 722–732. Authors from Cambridge University UK use a desk area network (DAN) to extend the use of ATM to the device and processor interconnect within a multimedia workstation using ATM cells as the basic unit of data transfer. They show that their prototype ATM workstation is capable of multimedia support but, of course, not as fast as optimized proprietary products.

Basch, B., et al., "VISTAnet Deployment and System Integration Experiences," *IEEE Journal on Selected Areas in Communications*, Vol. 12, No. 6, Aug. 1994, pp. 1097–1109. Description of a nightmare systems integration effort: turning up a multisite ATM network without the appropriate network management tools, using prototype equipment and software, and learning as they went. Interesting for its comments on HiPPI interfaces and cabling difficulties and the need for overlay networks. With a FETEX 150 ATM switch and up to SONET OC-48 (2,488 Gbps) transmission facilities, the maximum theoretical rate with HiPPI on a point-to-point link was 480–490 Mbps, and the application aimed for 80% of it or 420 Mbps.

Bellman, B., "ATM Edge Switches: Evaluating Features and Functionality," *Telecommunications*, Sept. 1995, pp. 29–34. A switch market-oriented article with a number of interesting aspects. Defines "edge" and "core" switches as components of a carrier's ATM network, and "enterprise," "backbone," and "workgroup" switches as part of the customer premises environment (but admits that "edge" switches often serve as service multiplexers on the customer premise and provide a signaling interface into the public network—the "brains," with many "feeds, speeds and features"). Sees FR transport as today's leading ATM application. Notes standards activities underway: ATM Forum's frame relay UNI (FUNI) and ATM inverse multiplexing (AIM) for 2–8 T1/E1s. In congestion control discussion, refers to "goodput" (!). Contrasts edge switches from six vendors and provides their World Wide Web addresses.

Bellman, J., (ed.), "ATM Practical Implementations Here and Now," Advertising Supplement to *Data Communications*, Feb. 1996, pp. A1–A12. Written by Brook Trail Research for 3Com, the several included articles contain various little nuggets. 3Com sees the desktop market dominated by price per port, the LAN market by scalability, and WAN market by bandwidth efficiency. Sees four "solution spaces:" desktop interconnect, LAN backbone, WAN access, and WAN transport. Predicts that ATM inverse multiplexing (AIMUX) will be finalized later in 1996, that QoS support will be defined in LANE 2.0, and that LANE 2.0 will be compatible with version 1.0 LECs. In network management, currently one must make do with the local management interface (ILMI), part of UNI 3.1, which uses SNMP. Relative to the future M1–M5 framework, M1 and M2 will use SNMP, where M5, between carriers, will use CMIP. On LANE, points out that "3Com's CELLplex 7000 ATM switch includes an i960 RISC processor that support LES, BUS, and LECS functions internally" and that LANE "adds insignificant overhead to internetwork traffic and barely makes a dent in the performance improvement obtained by deploying ATM in the local backbone." Further, on LANE performance, "frequent traffic between the two endpoints keeps the VC open and avoids additional overhead;" "direct VCs are torn down only after periods of inactivity" and "the busiest connections incur the least overhead." He goes on to observe that "with CELLplex 7000 ATM switches, LES and BUS transactions take less than 1 millisecond each; call setup takes less than 10 milliseconds per switch." In his example, with six components in the path there is less than 40 milliseconds before the direct VC is established, and with fewer hops call setup would be faster. Further, that a large percentage of the time, a direct VC would still be place from the previous transaction, making call setup unnecessary.

Berenbaum, A., et al., "A Flexible ATM-Host Interface for XUNET II," *IEEE Network*, July 1993, pp. 18–23. Describes an experimental high-speed (DS-3) network connecting several universities. Components included routers based on Silicon Graphics workstations and ATM switches based on the AT&T Hobbit RISC processors.

Beering, D. R., "Amoco Builds an ATM Pipeline," *Data Communications*, April 1995, pp. 112-120. An exciting account of Amoco's Aries ATM demonstration project. Using ATM equipment from FORE, General Datacomm, Newbridge, and Cisco, and T1/T3 links, it worked—sort of. Despite no SVCs, no end-to-end congestion control, no standard APIs, few high-speed ATM switch interfaces, and the lack of widespread availability of SONET, money, conservative practices, hard work and modest expectations triumphed.

Bennett, R. L., and G. E. Policello, II, "Switching Systems in the 21st Century," *IEEE Communications Magazine*, March 1993, pp. 24–28. Recounts the telephony shift from focusing on connecting fixed points, usually two, together rapidly, efficiently, and with high quality, to the brave new world of proliferating services, customer types, communications models, etc.

Berenhaum, A., et al., "A Flexible ATM-Host Interface for XUNET II," *IEEE Network*, July 1993, pp. 18–23. Describes an experimental high-speed network (@ DS-3) connecting several universities. Components include routers based on Silicon Graphics workstations and ATM switches based on the AT&T Hobbit RISC processor.

Beshai, M., R. Kositpaiboon, and J. Yan, "Interaction of Call Blocking and Cell Loss in an ATM Network," *IEEE Journal on Selected Areas in Communications*, Vol. 12, No. 6, Aug. 1994, pp. 1051–1058. Makes three points relative to the interaction between cell loss and call blocking under CBR and VBR. First, it is likely that operating link occupancy will be kept low to avoid blocking bursty traffic. Second, when CAC algorithms are used that guarantee x cell blocking with y cell loss, and they are set for full call level occupancy, and the cell occupancy is less than full, then the cell loss probability declines by two to three levels of magnitude. Third, tuning, "proper dimensioning," can help if the data requirements are well understood.

Biagioni, E., E. Cooper, and R. Sansom, "Designing a Practical ATM LAN," *IEEE Network*, July 1993, pp. 32–39. Describes FORE Systems' ATM switch, host interface cards, and the switch and interface card software. FORE's 16-port ASX-100 is aimed at the TCP/IP constituency and is intended to upgrade current LANs and/or provide a scalable backbone for campus networks.

Biersack, E. W., et al., "Gigabit Networking Research at Bellcore," *IEEE Network*, March 1992, pp. 42–48. An overview of Sunshine, Bellcore's ATM switch for AURORA. Describes the three classes of service—flow control, cell size, connection establishment—versus resource allocation; protocol data units (PDUs) versus packets; the TP++ transport protocol; ARC versus FEC; the three kinds of host interfaces; Batcher multi-Banyan switch fabric; and provides pertinent references.

Biersack, E. W., "Performance Evaluation of Forward Error Correction in an ATM Environment," *IEEE Journal on Selected Areas in Communications*, Vol. 11, No. 4, May 1993, pp. 631–640. Simulates the effectiveness of FEC for three traffic scenarios and finds that FEC reduces the loss rate for the video sources by several orders of magnitude for a heterogeneous traffic scenario consisting of video and burst sources.

Black, U. D., *ATM: Foundation for Broadband Networks*, Englewood Cliffs, NJ: Prentice Hall, 1995. Uyless Black's latest work views ATM in the wider concept of extant and future networks (indeed, only Chapters 6–11 and 13–15 are devoted to ATM). His description of ATM is basically the state-of-the-standard at the time of publication. Chapter 15 on "The ATM Market" identifies the major suppliers along with forecasts on the market and adaptor card costs.

Bogineni, K., K. M. Sivalingam, and P. W. Dowd, "Low-Complexity Multiple Access Protocols for Wavelength-Division Multiplexed Photonic Networks," *IEEE Journal on Selected Areas in Communications*, Vol. 11, No. 4, May 1993, pp. 590–604. Authors model and analyze two preallocation-based media access protocols for a wavelength-division, multiple-access, star-coupled photonic network.

Bonatti, M., et al., "B-ISDN Economical Evaluation in Metropolitan Areas," *IEEE Journal on Selected Areas in Communications*, Vol. 10, No. 9, Dec. 1992, pp. 1489–1502. A highly abstract economic/technical analysis of what it would cost to provide a 2-Mbps ATM service to a metro area of two million inhabitants; a centralized PTT-planner type perspective unrelated to market needs or demands.

Bonomi, F., and K. W. Fendick, "The Rate-Based Flow Control Framework for the Available Bit Rate Service," *IEEE Network*, March/April

1995, pp. 25–39. A history of the various congestion avoidance proposals set forth for ATM's ABR. The winner, a rate-based protocol, employs at its simplest a one-bit EFCI end-to-end mechanism in conjunction with a closed-loop feedback loop using a CI bit in the RM cells; concerns for NIC simplicity and low overhead at high data rates were persuasive. Optionally, though, intermediate switches or networks can segment the control loop, and switches can provide more detailed feedback that dynamically changes an explicit upper bound on the source end system (SES).

Bourne, J., and I. Cunningham, "Implementing the Information Highway," *Telesis*, No. 98, May 1994, pp. 4–25. Introduction to the capabilities and rationale behind the NT Magellan line of ATM products, Interesting is that in network management, they offer proprietary and SNMP-based systems, the latter touted as "open."

Brunet, C. J., "Hybridizing the Local Loop," *IEEE Spectrum*, June 1994, pp. 28–32. Presents the amazing variety of hybrid copper/coax/fiber alternatives being forwarded by the cable/telco/energy companies. Connection-oriented, telcos today are logical stars and point to point, cable companies are logical buses and point to multipoint. But convergence...?

Bruwer, W. A., et al., "VISTAnet and MICA: Medical Applications Leading to the NCIH," *IEEE Network*, Vol. 8, No. 6, Nov./Dec. 1994, pp. 24–31. Describes an ATM prototype in North Carolina, coupling supercomputers with a high-speed network for radiation therapy. With prototype equipment, draft standards, and no performance or network management capability, a nightmare to turn up. Used Fujitsu FETEX 150s and PVCs with HiPPI interfaces. Network at 622 Mbps ran at 480 Mbps (23% overhead) once SONET, ATM cell, and AAL 3/4 protocols weighed in.

Bryenton, A., and S. Shew, "A Powerful Architecture for Enterprise Networking," *Telesis*, No. 98, May 1994, pp. 69–78. Describes the features of NT's Passport ATM switch. Believes that organizations will use them to build one big network ("single enterprise backbone"); that they will become deployed as "infrastructure" rather than "overlay networks;" that they will use the discard data feature to avoid congestion rather than "over-engineering."

Chiang, A., "Parallel Paths Emerge for Fast Ethernet and ATM," *Telecommunications*, March 1996, pp. 38–39. Sees 100 BaseT and ATM as complementary, the former to move data faster, the latter to move multimedia. Too, he sees 100 BaseT at the workstation, and ATM at the network edge and on the corporate backbone.

Capell, R. L., et al., "Evolution of the North Carolina Information High-way," *IEEE Network*, Nov./Dec. 1994, pp. 64–70. Amusing marketing article in which Bell South tries to flatter North Carolina politicos into "one big [ATM] network" from guess who. Amusing, too, in that the technical features in the marketing blurb do not match those in the technical articles. Clear that Bell South sees in ATM the "killer technology" to recapture functionalities now provided by customer-owned CPE.

Chao, H. J., "A Recursive Modular Terabit/Second ATM Switch," *IEEE Journal on Selected Areas in Communications*, Vol. 9, No. 8, Oct. 1991, pp. 1161–1172. Basically, an argument for a switch design, one that, within current CMOS technology, would combine (and scale up) a number of 155-Mbps channels over 8,192 input ports to an aggregate switch throughput rate of greater that 1 terabit/s. Uses crossbar fabric with SWEs, GNs, and MIMO muxes to achieve a tightly coupled line concentrator/switch fabric design. In introduction, argues that NGs can be characterized by buffer placement.

Chao, H. J., et al., "IP on ATM Local Area Networks," *IEEE Communications Magazine*, Aug. 1994, pp. 52–59. The problem discussed is how to support "connectionless" broadcast LANs using IP as their network protocol via ATM, a connection-oriented point-to-point service. Two approaches are forwarded for the PVC case: LAN emulation and IP over ATM. In the former, ATM is made to look and behave like an 802 MAC protocol below LLC. In the latter, there is IP to ATM address resolution. Both require high-speed address services, which could become performance choke points and introduce implementation complexity.

Chen, T. M., and S. S. Liu, *ATM Switching Systems*, Norwood, MA: Artech House, 1995. Written by two electrical engineering Ph.D.s from GTE, Chen and Liu concentrate on developing what they call the "general ATM switching system model," with an emphasis on switching and network control. Highly detailed, they assume a SONET physical layer and organize their discussion along five "functional areas:" input, output, cell switch fabric, connection admission control, and system management.

Chen, T. M., and S. S. Liu, "Management and Control Functions in ATM Switching Systems," *IEEE Network*, July/Aug. 1994, pp. 27–40. Authors argue that ATM management and control functions will be significantly more complex in the building of an ATM switch than the switch fabric. This is due to the fact that usually within a distributed architecture, control must be exerted at the levels of every VC and at/with individual cells.

Chipalkatti, R., Z. Zhang, and A. S. Acampora, "Protocols for Optical Star-Coupler Network Using WDM: Performance and Complexity Study," *IEEE Journal on Selected Areas in Communications*, Vol. 11, No. 4, May 1993, pp. 579–589. The environment/test givens are that fiber bandwidth is restricted by digital electronics, thus how to optimize a passive star coupler with wavelength division multiplexing using a fixed transmitter and tunable receivers. Authors compare performance of DAS, TDM, and hybrid TDM. Conclusions are that TDM is best for WAN, DAS and HTDM for LAN environments. DAS has the highest performance but high signaling costs limit it to small nets. HTDM is better than TDM for mid-sized nets and bursty traffic. In general, reservation-based schemes with no collisions are superior for heavy loads and multiple access schemes with collisions are superior for low loads.

Chlamtac, I., "An Optical Switch Architecture for Manhattan Networks," *IEEE Journal on Selected Areas in Communications*, Vol. 11, No. 4, May 1993, pp. 550-559. Proposes a new electronically controlled optical packet deflection switch based on space and time switching at intermediate nodes.

Cisneros, A., and C. A. Brackett, "A Large ATM Switch Based on Memory Switches and Optical Star Couplers," *IEEE Journal on Selected Areas in Communications*, Vol. 9, No. 8, Oct. 1991, pp. 1348–1360. Authors present an experimental prototype switch design that combines electronic information processing and switch control with multiwavelength optical interconnection networks, a particular example which, in theory, will achieve a switch capacity of 2.5 Tbps. Claim that the signal processing and contention resolution are within the capability of present-day electronics for a switch with 16,384 STS-3c input lines and that the optical interconnect network is feasible, although not yet practical. Proclaims (p. 1348) that "the possibility of purely photonic packet switching is (at present) severely limited by the difficulty of buffering packets and reading packet headers in the photonic domain. The buffering function is necessary in a packet switch unless it is unrealistically overbuilt. This function can only, at present, be provided electronically in the needed capacity."

Cohen, J., "Sun Beams ATM Traffic At Higher Speed," *Network World*, April 15, 1996, pp. 1, 57. News article announcing Sun's 622 (OC-12) Mbps Sbus multimode fiber NIC for $4,995.

Cohen, J., "ATM Under the Gun," Network World, May 13, 1996, pp. 1, 65. News article sees ATM's acceptance slowed by rival technologies in key market areas, such as 100-Mbps Ethernet in the desktop market and switched fast Ethernet in the LAN workgroup market. Also sees evolutionary LAN technology offsetting ATM advantages: higher speeds by fast Ethernet or gigabit Ethernet; dedicated bandwidth by switched Ethernet; isochronous support by isochronous Ethernet, IEEE 802.9, or 3Com's PACE technology; better management by virtual LANs; and multicast and bandwidth reservation by enhanced protocols such as Ipng and RSVP.

Cooney, M., "ATM Forum Looks to Solidify Market," *Network World*, April 29, 1996, p. 9. A news article elaborating on the ATM Forum's recent "Anchorage Accord," freezing most new standards development for the next two years. The accord is an attempt by the Forum to stabilize standards so as to increase the longevity of ATM products and consists of "foundation layer" specifications and "extensions" to be built on top of them. The foundation layer components include the Physical Layer specification, Interim Layer Management Interface 4.0, Traffic Management 4.0, Signaling 4.0, Private Network-to-Network Interface 1.0, Broadband Inter-Carrier Interface and Network Management. Extensions will include Multi-Protocol over ATM (MOPA) and LAN Emulation (LANE) 1.0. Reports further speculation on whether the new ABR specification, part of Traffic Management 4.0, will require "forklift" upgrades of present switch and NIC cards.

Cox, J. R., Jr., M. E. Gaddis, and J. S. Turner, "Project Zeus," *IEEE Network*, March 1993, pp. 20–30. Account of an ATM-based campus network at Washington University at St. Louis. Initial deployment used homogeneous, customized, 16-port, 155-Mbps SynOptics switches.

Crutcher, L. A., and A. G. Waters, "Connection Management for an ATM Network," *IEEE Network*, Nov. 1992, pp. 42-55. Authors contend that the existing techniques for connection management do not adequately meet the requirements of applications that are envisioned for ATM networks and that a reworking to allow multiservice and multiparty communication is required. They believe that these problems will involve significant complexity and that the processing overhead for the traffic in the network related to management and control will consume a significant portion of the network resources.

Csenger, M., "ATM Hums, But Can It Sing?," *Network World*, Oct. 3, 1994, pp. 1, 92. Article points out that supporting voice with AAL1 is essentially a nailed-up circuit with 18% overhead and argues the need for an efficient SVC-oriented way to handle voice via ATM over DS-3.

Cuthbert, L. G., and J. C. Sapanel, *ATM, The Broadband Telecommunications Solution*, London: The Institute of Electrical Engineers, 1993. This overview of ATM principles and issues is a group effort of project "R1022: Technology for ATD," part of the RACE program of the Commission of the European Communities.

Dagdeviren, N., et al., "Global Networking With ISDN," *IEEE Communications Magazine*, June 1994, pp. 26–32. States the promise of ISDN is not speed or performance but in universal connectivity; with its with sophisticated signaling and control, it offers a well-defined evolutionary path for the PSTN.

Damodaram, R., et al., "Network Management for the NCIH, *"IEEE Network*, Nov./Dec. 1994, pp. 48–54. For NCIH OA&M, Bell South chose an overlay structure to manage a multicarrier, multivendor ATM network. At the center was the broadband overlay operations systems (BOSS). BOSS uses a Sun Sparc with Solaris, Informix and Informix 4GL, C-PLEX Integer Problem Solver, and NetExpert gateway to connect with the many players and manage the resources.

Robert, P., and N. J. Muller, *The Guide to Frame Relay and Fast Packet Networking*, New York, NY: Telecom Library, 1991. A nice introduction to frame relay (FR) that raises interesting questions about ATM. Since FR is farther along in the standards game and is capable of high speeds (T-3 today), and ATM is basically being pushed by the data side of the house, is the CBR constituency (video mostly) important enough to justify ATM over FR? As ATM overlaps FR, but is less efficient for computer data, are they likely to coexist?

Davidson, R. P., and N. J. Muller, *The Guide to SONET, Planning, Installing & Maintaining Broadband Networks*, New York, NY: Telecom Library, 1991. This is a useful book on the synchronous optical network (SONET) transport standard. It contains as well a thorough discussion of T-1/T-3 network technology. It is understandably theoretic as SONET was in the early trial stage and there were no large-capacity switching devices available that were designed for SONET.

de Prycker, M., *Asynchronous Transfer Mode, Solution for Broadband ISDN*, New York, NY: Ellis Horwood, 1991. An excellent but fairly difficult early work describing the thinking behind the ATM standard; given its publication date, it does not deal with ATM implementations.

de Prycker, M., "ATM Switching on Demand," *IEEE Network*, March 1992, pp. 25–28. Raises a number of points on just how ATM will be implemented, virtual channel connections (VCC), and virtual path connections (VPC), and possibly VCCs within a VPC. Speculates that to keep multiplex costs down, service muxes may be simple (transparent data exchange) but may provide a number of QoS functions; too, centralized servers ("connectionless" servers) may be used to keep service mux costs down. Nor is it clear how multimedia will be handled, as a "big pipe" application or as many small applications and reassembled. Stresses the importance of the upcoming CCITT signaling standard relative to the above issues.

de Prycker, M., R. Peschi, and T. Van Landegem, "B-ISDN and the OSI Protocol Reference Model," *IEEE Network*, March 1993, pp. 10–18. Maps the ATM/B-ISDN protocol reference model to OSI and concludes that the ATM layer is equivalent to the OSI physical layer and the service offered by AAL 3/4 is equivalent to the OSI data link layer. Many useful tidbits on the history of ATM, why it is "asynchronous," its three layers (ATM physical for transport, ATM layer for switching and muxing, and AAL for adapting service information into the ATM stream), the elements of AAL (SAR, CS), the origins of SEAL, and how ATM uses separate VC's to implement out-of-band signaling.

Decina, M., C. Mossotto, and A. Roveri, "The ATM Test-Bed: An Experimental Platform for Broadband Communications," *IEEE Communications Magazine*, Oct. 1994, pp. 78–83. The article describes a Milan-area ATM lab designed to test ATM-based internetworking between heterogeneous devices. The various equipment technologies are described in some detail. The experience showed the difficulties in proving in different versions of protocols with a total lack of network management and control facilities.

Denzel, W. E., "High-Speed ATM Switching," *IEEE Communications Magazine*, Vol. 31, No. 2, Feb. 1993, p. 26. Observes (1) that ATM is being introduced first in private networks; (2) sees ATM bringing a "packetized, unified, broadband telecommunications and data communications world;" (3) that ATM is "a universal, service-independent switching and multiplexing technique...totally independent of the underlying transmission technology and speed;" and (4) in the LAN market predicts LAN-like services "emulated by switches."

Dhas, C., V. K. Konangi, and M. Sreetharan, (eds.), *Broadband Switching: Architectures, Protocols, Design, and Analysis*, Los Alamitos, CA: IEEE

Computer Society Press, 1991. Some 47 articles, most reprinted from IEEE publications, under the headings of "Network Architecture," "Interconnection Networks," "Experimental Architectures," "Switch Fabric Design and Analysis," "Switch Architectures," "Broadband Switching Networks," "Bandwidth Allocation, Flow Control, and Congestion Control," "Performance Modeling," and "Photonic Switching Systems," Brief introductions precede each topic; most of the ATM material is in the "Experimental Architectures" section.

Drynan, D., and J. Ellis, "A New Class of Switch for the Multimedia Age," *Telesis*, No. 98, May 1994, pp. 53–67. A fairly high-level description of NT's "edge" switch for ATM, the Magellan Passport.

Editors, "Gigabit Network Testbeds," *Computer*, Sept. 1990, pp. 77–80. A brief description, as of Sept. 1990, of the Corporation for National Research Initiatives' gigabit testbeds, the organizations involved, the computer science and application research areas, and the proposed hardware.

Editors, "Packet and Cell Switching," *Data Communications*, Oct. 21, 1993, pp. 131–137. Article is part of magazine's 1994 Internetworking Product Guide Issue and divides the ATM switch world into LAN and campus switches (most under $50K, and oriented toward the LAN environment) and private networking switches ($100K and up, with greater capacity and offering a variety of WAN interfaces). Offers brief overviews of 26 products from a somewhat fewer number of vendors.

Editors, "Telecommunications," *The Economist*, Sept. 30, 1995, pp. 1–29. A far-ranging—and exciting—view of the world telecommunications future. Their thesis is that "the death of distance as a determinant of the cost of communications will probably be the single most important economic force shaping society in the first half of the next century."

English, E., "Ethernet Looks to a Gigabit Future," *Computer*, June 1996, p. 18. News article tracks the development of the Gigabit Ethernet Alliance, which hopes to have a spec by late 1998, and a parallel effort by the IEEE on gigabit VG-AnyLAN. Cites Dataquest on the Ethernet installed base being 70M, and that Fast Ethernet (100Base-T Ethernet) will reach 4.2M, and 100VG-AnyLAN 780K, in 1996; by 1997, the predictions are 10.6M for fast Ethernet, 1.7M for 100VG-AnyLAN.

Feldmeier, D., "A Framework of Architectural Concepts for High-Speed Communications systems," *IEEE Journal on Selected Areas in Communications*, Vol. 11, No. 4, May 1993, pp. 480–488. He overviews a number of recent papers concerned with the upper protocol layers (ISO layer 4 and above) "because the upper protocol layers are seen to be performance bottlenecks." (p. 480) Among these difficulties, he sees (1) multiplexing control and data (as in datagrams); (2) providing more services than needed and/or duplicating network-provided services; (3) the conflicts between ordering for error detection, encryption, and/or data compression and the efficiencies promised by "pipelining among layers ("structural parallelism") and pipelining within layers ("pipelining blocks")" (p. 484).

Fischer, W., et al., "A Scale ATM Switching System Architecture," *IEEE Journal on Selected Areas in Communications*, Vol. 9, No. 8, Oct. 1991, pp. 1299–1307. Describes a Siemens prototype ATM switch. Perhaps the most interesting feature is its approach to fault tolerance, which provides for each VC/VP a parallel redundant path without synchronization of parallel switch planes.

Fischer, W., et al., "Data Communications Using ATM: Architectures, Protocols, and Resource Management," *IEEE Communications Magazine*, Aug. 1994, pp. 24–33. Excellent, though difficult, article that discusses data uses of ATM and the challenges they present for protocol processing and resources management. Among the aspects discussed are the ATM standards effort, the idea of computers with ATM output channels, fat versus skinny protocols, multicast error handling, TCP over ATM, buffer sizes, multiplex buffers and "egress shaping," the delay-bandwidth product, congestion controls with large round-trip delays, and resource management.

Flanagan, W. A., *Frames, Packets and Cells in Broadband Networking*, New York, NY: Telecom Library, 1991. A well-conceived book whose perspective is the nuts and bolts of information segmentation (frames, packets, cells). It contains a chapter on each of the broadband protocols; it does not address actual networks or product implementations.

Forkish, R., *Multiservice Backbone Networks*, Redwood City, CA: Network Equipment Technologies, 1994. In the context of an overview of multiservice backbone networks from a smart multiplexer supplier, a number of interesting observations on ATM and transitioning to it.

Fotedar, S., et al., "ATM Virtual Private Networks," *Communications of the ACM*, Vol. 38, No. 2, Feb. 1995, pp. 101–109. Typical academic piece modeling to great detail the unknowable. But the basic idea is sound: usage parameter control (UPC), or policing, or input shaping is hard to do, and using VPs to create virtual subnets that permit subdivision of the traffic into more homogeneous, more manageable groups allows one, at the cost of some statistical muxing efficiency, to better control the situation.

Fowler, H. J., "TMN-Based Broadband ATM Network Management," *IEEE Communications Magazine*, March 1995, pp. 74–79. Disturbing article on telecommunication management network (TMN) for ATM. ATM Forum's plan parallels the five-layer TMN hierarchy (EL, EML, NML, SML, BML) with its M1–M5 levels but is defining protocol-independent MIBs given the conflict between the computer/private network constituency that is SNMP-oriented and the telecom/public network constituency that is CMIP-oriented. Given that the usual networking combinations, public-public, public-private, private-private, are necessary to make ATM work, it bodes ill.

Franz, R., K. D. Gradischnig, M. N. Huber, and R. Stiefel, " ATM-Based Signaling Network Topics on Reliability and Performance," *IEEE Journal on Selected Areas in Communications*, Vol. 12, No. 3, April 1994, pp. 517–525. In an acronym-clogged article, the authors overview and speculate on how ATM will be integrated within today's SS7 networks. First, there are the competing upgrade versus overlay implementation scenarios. Second, signaling traffic is bursty. Current SS7/STM networks use 64K bands for signaling with mean message lengths of 100 bytes. Future ATM networks, because of growth in features and signaling relations and dependencies, will use VCCs of up to 4K bytes at 1 Mbps. Current requirements for unavailability run less than 10 minutes/year. This should be achievable in the ATM environment with $1 + 1$ or N in M hardware backup.

Frymoyer, E. M., "Fiber Channel Fusion: Low Latency, High Speed," *Data Communications*, Feb. 1995, pp. 107–112. Fiber Channel, an ANSI standard, is a variable-media, variable-topology channel extension technology (up to 10 km with single-mode fiber) that offers QoS, high speed (to 800 Mbps), and variable frame sizes for high efficiency. Less certain will be its popularity outside of the supercomputer environment.

Fujii, H., and N. Yoshikai, "Restoration Message Transfer Mechanism and Restoration Characteristics of Double-Search Self-Healing ATM Network," *IEEE Journal on Selected Areas in Communications*, Vol. 12, No. 1, Jan. 1994, pp. 149–158. Advocates virtual path (VP) protection switching for ATM networks, but using a novel double-search (bidirectional) self-healing algorithm. Although a simulation with 110 nodes confirms its effectiveness, less than 100-ms restoration (average path restoral in Japan today is 10 seconds!), the cells for the restoration information are different from the OAM cells recommended by current international standards. Interestingly, study assumes mean link use at 48% (SDH) path capacity.

Gage, B., "ATM Under the Covers," *Network World*, July 10, 1995, pp. 45–46. Overview of what the carriers are offering in ATM services (CBR, VBR, with some ABR), interfaces (DS3, OC-3, some DS1), and the switches they are using (AT&T GlobeView-2000, GDC Apex-NPX, Strata-Com BRX, Newbridge 36150, and NEC NEAX-61E and Model 10).

Gage, B., "' Frame Relay, Meet ATM,'" *Network World*, Nov. 6, 1995, pp. 57–60. Relates the approaches taken by vendors (AT&T, Cascade, GDC, Newbridge, Nortel, StrataCom) to FR-to-ATM interfacing. Distinguishes between "network interworking" where one protocol is used on both ends and another in between, and "service interworking" where there is proto-col conversion and feature-to-feature mapping is involved. The Forum's FUNI is the coming thing in service interworking as it offers SVC support. Sprint, today, is the only vendor offering commercial FR-to-ATM service and they use service interworking to provide a 0 CIR (!) PVC.

Garcia-Haro, J., and A. Jajszczyk, "ATM Shared-Memory Switching Architectures," *IEEE Network*, July/Aug. 1994, pp. 18–26. Argues for shared-memory-based ATM switch designs and discusses a number of de-sign variations as well as their advantages and disadvantages.

Gareiss, R., "An ATM Service for Wide-Open Spaces," *Data Communica-tions*, March 21, 1995, pp. 43–44. News article. US West, using General Datacomm Apex ATM switches, offers CBR and VBR services to 155 Mbps at 40–50% less than leased line alternatives.

Gelekas, J., and B. Bressler, "ATM to the Desktop: 25 Mbit/s or 155 Mbit/s?," *Data Communications*, Sept. 1995, pp. 101–106. Interesting tid-bits contained in an article contrasting 25 Mbps versus 155 Mbps for desktop ATM. IBM argues cost, especially in less expensive buffers in the NICs and switch ports; Sun argues future bandwidth needs, 2x cost for 6x performance. Tidbits: "playout buffers" at the client mitigate jitter prob-lems that isochronous service avoids; IBM arguing for packet transfer mode (PTM) with variable-length cells for more efficient WAN transfers; IETF's RSVP (resource reservation protocol), making it possible for layer 3 router-based networks to guarantee QoS support similar to that of ATM.

Giacopelli, J. N., et al., "Sunshine: A High-Performance Self-Routing Broadband Packet Switch Architecture," *IEEE Journal on Selected Areas in Communications*, Vol. 9, No. 8, Oct. 1991, pp. 1289–1298. Authors present a self-routing broadband packet switch architecture based on Batcher-Banyan networks that combines a recirculating queue with output queues in a single architecture. Authors claims that output queues, when combined with recirculators, shift the queuing burden away from the recirculators, enabling the shared queue to effectively control momentary output overloads and enabling the architecture to achieve the extremely low cell-loss probabilities that may be required by circuit emulation services. Further advocates that where large aggregates of traffic flow between locations, trunk grouping may be used to form high-capacity channels. These channels use multiple transmission links and serve as one high-bandwidth pipe. Logical addressing is used to decouple switch-specific control information from the external addressing information used by the cells and to form trunk groups of arbitrary sizes out of arbitrary sets of transmission links.

Gopal, I., et al., "ATM Support in a Transparent Network," *IEEE Network*, Nov. 1992, pp. 62–68. An important article that argues that the ATM cellular approach has a number of unattractive, pushed-to-the-network-periphery side effects: the creation of considerable adaptation layer processing, what used to be called protocol conversion, in order to get simplicity and high speed. (This is very similar to today's digital hierarchy and its 64-Kbps TDM slotting.) With the plaNET switch, a hybrid approach, there is a frame relay switch providing transparency (minimal adaptation processing), and ATM native mode support. Claims hybrid will be more efficient (re: AUROA testbed for the numbers) and will allow cheaper workstation adaptors and less adaptation layer processing. Interestingly, plaNET supports three routing schemes concurrently: source routing (the preferred), node-to-node (like ARPAnet), and "pathfinding." Authors argue they can do so (frame relay cum ATM, three kinds of routing support) with little additional switch complexity.

Goralski, W. J., *Introduction to ATM Networking*, New York, NY: McGraw-Hill, 1995. Written by a member of the ATM Forum, it basically explains the state of the ATM standard in late 1994, but makes it clear that the standard is very much a work in progress. The final chapter contains a brief survey of ATM and network service providers.

Green, D., *Telecommunications*, March 1996, pp. 36–37. A pitch for ATM edge routers from NetEdge Systems. Sees three ways to handle legacy LANs in an ATM network: LAN switches with ATM connectivity, routers with ATM plug-in cards, and edge routers. Argues the latter perform best because they allow huge packet sizes (to 64 KB with IP) to optimize server performance.

Grovenstein, L. W., et al., "NCIH Services, Architecture, and Implementation," *IEEE Network*, Nov./Dec. 1994, pp. 18–22. The NCIH is, in late 1994, probably the world's largest ATM network with twelve ATM switches and seven ATM concentrators. They use (currently) circuit emulation for distance learning at DS-3 (.02% additional overhead with AAL 1) and SMDS for packet-switching datagrams (8% additional overhead with AAL 3/4).

Guarneri, R., and C.J.M. Lanting, "Frame Relaying As A Common Access to N-ISDN and B-ISDN Data Services," *IEEE Communications Magazine*, June 1994, pp. 39–43. Argues for frame relay services, mainly for its potential to provide low cost per kilobyte service, over N-ISDN and B-ISDN.

Guruge, A., "Can ATM Rescue IBM?," *Network World*, June 5, 1995, p. 45–48. Argues that IBM does have a unique, full-spectrum ATM product range, with a "value-enhancing architecture" reminiscent of SNA. Further, after it's lapse in multiprotocol router internetworking, this potential for a single-vendor solution could give IBM an edge to succeed with ATM.

Haas, Z., et al., "Guest Editorial: Protocols for Gigabit Networks," *IEEE Journal on Selected Areas in Communications*, Vol. 11, No. 4, May 1993, pp. 477–479. In their introduction to the special issue on high-speed protocols, they define the objective as gigabits end to end; point out that the higher one goes in the ISO protocol hierarchy, the greater the uncertainty; and report general skepticism on the appropriateness of the ISO hierarchy generally.

Habib, I. W., "Neurocomputing in High-Speed Networks," *IEEE Communications Magazine*, Oct. 1995, pp. 38–40. Habib stresses the potential benefits of applying neurocomputing to solve some challenging engineering problems in high-speed networks, including multimedia traffic characterization, adaptive allocation of resources, flow and congestion control, maintenance of QoS, and dynamic routing.

Händel, R., M. N. Huber, and S. Schröder, *ATM Networks, Concepts, Protocols, Applications*, New York, NY: Addison-Wesley, 1994. The authors, from the Public Communications Networks Group of Siemens in Munich, reflect their strong standards background (ITU-T, ETS1, T1, ATM Forum) in a conventionally organized introduction to ATM, with many short chapters covering B-ISDN, standards, signaling, ATM switching types, and transmission. Most interesting is Craig Partridge's out-loud wondering about ATM as a "coherent link level protocol" and how in modeling ATM performance there are competing models, drawn respectively from the telephony and data worlds, that of "big bits" (bigger TDM units) versus "small packets."

Havens, R., and M. Sievers, "Numbering in a Competitive Environment," *IEEE Communications Magazine*, July 1994, pp. 68–70. Argues that the administration of the North American Numbering Plan (NANP) for the PSTN was left in control of the LECs, and they have used it to hinder the introduction of new and competitive services. Numbers are like bandwidth, and with the coming of the IN, are too important to be left to the LECs.

Herman, J., and C. Serjak, "ATM Switches and Hubs Lead the Way to a New Era of Switched Internetworks," *Data Communications*, March 1993, pp. 69–84. Sees current LANs/Internets running out of capacity and unable to handle new image applications. Sees the future for data networks to be "switched Internets" when dedicated Ethernets are hubbed to ATM switches, especially in the campus environment. Overviews the ATM plans of major router/hub vendors.

Heywood, P., "Can ATM Really Be This Cheap?," *Data Communications*, Jan. 1995, p. 71. A news story on how Telecom Finland Ltd. is the first non-U.S. public network operator to offer long-distance ATM service and how its prices are an order of magnitude lower than its U.S. competitors.

Hold, D. F., (ed.), "Performance Benchmarks," *The ATM Report*, Jan. 31, 1996, p. 3. Cites a fall, 1995 *Network Computing* study of performance using 155-Mbps PC adaptors. EISA only got one-third of the 155, where PCI cards max out at 100 Mbps. Restraints are small packets, TCP rather than UDP, and workstation disk systems. Sees huge variations in performance and many ways to unrealistically juice the numbers. Issue also contains (pp. 4–10) an excellent guide to ATM NICs.

Hold, D. F., (ed.), "The ATM Report's Internet ATM Survey," *The ATM Report*, Sept. 23, 1996, pp. 2–11. Survey of 20 providers of Internet services relative to plans for ATM-provided backbone and access. Most have a "need for speed" and ATM is the only commercial alternative at OC-3 and above. Other reasons are the quick response available with explicit rate (ER) flow control with ABR. As supporting the current customer base is paramount, obligatory is a mix of IP, FR, and ATM. The biggest downside of ATM continues to be IP support. Current Internet traffic is many short IP packets, 55% 64 bytes, 25% 32 bytes. A 64-byte IP packet takes two ATM cells, 106 bytes, to transport. Also needed is RSVP-to-ATM QoS mapping. RSVP has the further problems of reservation on a per-port basis and no mechanism with which to pass RSVP requests across networks.

Horwitt, E., "IP Over ATM," *Network World*, April 15, 1996, pp. 38–45. A survey article on IP-over-ATM pitfalls. Included are the three choices (classical IP, LANE, and PVCs), the first two generating router bottlenecks, the latter the necessity for static router table administration. Mentioned are such nonstandard approaches as FORE's Simple Protocol for ATM Network Signaling (SPANS)—fine on the LAN, but it doesn't scale for the WAN—and Ipsilon Networks' IP Switch ATM 1600, which doesn't connect with other ATM switches. MOPA is a future item and currently the subject of a Bay versus Cisco power struggle. No approach supports QoS, so extra capacity is a must as IP bursts may knock down SNA sessions. Mentions IBM's plans to map SNA QoS, and IETF's RSVP, to ATM QoS. Ends with the observation that IP networks have historically ignored network engineering, a sure recipe for disaster with ATM networks.

Hoshi, T., et al., "B-ISDN Multimedia Communication and Collaboration Platform Using Advanced Video Workstations to Support Cooperative Work," *IEEE Journal on Selected Areas in Communications*, Vol. 10, No. 9, Dec. 1992, pp. 1403–1412. Describes a Hitachi protoype lab to advance distributed, workstation-based, collaborative work employing audio and video teleconferencing and shared object manipulation. The environment is linked using ATM/ BISDN at 156 Mbps.

Houh, H. H., et al., "The VuNet Desk Area Network: Architecture, Implementation, and Experience," *IEEE Journal on Selected Areas in Communications*, Vol. 13, No. 4, May 1995, pp. 710–721. Describes a prototype desk area network (DAN) called VuNet. Within a small network "trust-boundary," a simple, high-speed ATM LAN was built in which storage and video devices were directly connected to the net with custom software interfaces as opposed to workstations. Workstations, and their poor I/O, they find, are an impediment to high speed and flexible multimedia.

Hughes, D., and K. Hooshmand, "ABR Stretches ATM Network Resources," *Data Communications*, April 1995, pp. 123–128. An interesting article by StrataCom representatives that details the service class hierarchy (CBR, VBR [VBR/RT, VBR/NRT], ABR, and UBR) and the rate-based, closed-loop congestion control approaches (EFCI, ERM, segmented VS/VD, hop-by-hop VS/VD) available to switch manufacturers.

Hughes, J. P., and W. R. Franta, "Geographic Extension of HiPPI Channels via High Speed SONET," *IEEE Network*, May/June 1994, pp. 42–53. The subject is a fiber version, employing SONET, to extend the copper-based HiPPI channel. The objective is to allow one or more supercomputer transfers at 800 Mbps (or in such aggregates) over fiber facilities up to 3,000 miles apart. The article describes both a prototype product designed and built by NSC and a draft standard for HiPPI-SONET operation. The article also describes parallel activities in the Casa and Nectar groups including HiPPI-ATM-SONET.

Hui, J. Y., J. Zhang, and J. Li, "Quality-of-Service control in GRAMS for ATM Local Area Network, *IEEE Journal on Selected Areas in Communications*, Vol. 13, No. 4, May 1995, pp. 700–709. Describes the Rutgers Gopher-style real-time ATM multimedia services (GRAMS) testbed for video, image, and text services. A number of experiments are performed to measure QoS and throughput. (Peculiarly, "starvation" in the article refers to underflow.) Finds out in throughput studies that ATM AAL 4 using FORE API and SPANS doubles the throughput over TCP/IP over ATM and that new disk technology is essential to improved system performance.

Humblet, P. A., R. Ramaswami, and K. N. Sivarajan, "An Efficient Communication Protocol for High-Speed Packet-Switched Multichannel Networks," *IEEE Journal on Selected Areas in Communications*, Vol. 11, No. 4, May 1993, pp. 568–578. In an attempt to integrate transport layer and media access layer functions, paper proposes a new media access protocol for high-speed, packet-switched, multichannel networks based on a broadcast topology.

Hyman, J. M., A. A. Lazar, and G. Pacifici, "A Separation Principle Between Scheduling and Admission Control for Broadband Switching," *IEEE Journal on Selected Areas in Communications*, Vol. 11, No. 4, May 1993, pp. 605–616. This article describes a performance simulation employing the principle of separation and an admission control strategy designed to maximize the expected system utility while maintaining all QoS guarantees.

de la Iglesia, R. D., "Parametric Methodology for Analyzing Implementation Strategies of Residential Broadband Networks: The Incremental Cost of Integrated Access," *IEEE Journal on Selected Areas in Communications*, Vol. 10, No. 9, Dec. 1992, pp. 1510–1522. The paper describes a methodological framework based upon engineering modeling for comparing the investment cost of two broadband implementation strategies (namely dual versus single network) for residential customers and concludes that an integrated approach is both more expensive and, for several reasons, less likely.

Inoue, Y., et al., "Granulated Broadband Network Applicable to B-ISDN and PSTN Services," *IEEE Journal on Selected Areas in Communications*, Vol. 10, No. 9, Dec. 1992, pp. 11474–1488. Argues for a granulated broadband network (GBN) that accelerates "ATMization" via aggressively shifting current PSN traffic to B-ISDN via single-channel, cell assembly-disassembly devices (CLAO). Describes the ATM service concept with a clever football ticket metaphor. Sees ATMization as the "second evolution" of the PSN after digitization. Criticizes STM's mux and multistage transmission hierarchy for inefficient transit link utilization; says ATM requires less transit link capacity and yields higher utilization.

Inoue, Y., and N. Terada, "Granulated Broadband Networks," *IEEE Communications Magazine*, April 1994, pp. 56–62. Authors, from NTT and drawing on a Japanese setting, pitch ATM's ability to provide granulated services (bandwidth × % call blockage × % cell loss) and believe it is (or will be) economical even at the low end, the home/small business market. And that this will happen even with the costs of installing fiber to the residence/business and without broadband video on demand (strictly regulated by the government if provided by the telephone operating companies).

Itoh, A., "A Fault-Tolerant Switching Network for B-ISDN," *IEEE Journal on Selected Areas in Communications*, Vol. 9, No. 8, Oct. 1991, pp. 1218–1226. Paper proposes a self-routing fault-tolerant switching network for ATM switching systems that employs many "subswitches" to enhance the fault tolerance of the conventional multistage interconnection network, which only has a unique path. The subswitches provide large numbers of alternative paths between switching stages and allow the network to tolerate multiple faults.

Itoh, A., et al., "Practical Implementation and Packaging Technologies for a Large-Scale ATM Switching System," *IEEE Journal on Selected Areas in Communications*, Vol. 9, No. 8, Oct. 1991, pp. 1280–1288. The authors describe NEC's prototype ATOM (ATM output buffer modular) switch with 8 × 8 310 Mbps per port modules, that in a three-stage structure with 48 modules results in a 64 × 64 switch with 156-Mbps ports. Authors believe that the ATM switch of the late 1990s should support 1,000 156-Mbps ports and that all ATM switches are of four types: input buffer, shared buffer, output buffer, and crosspoint buffer.

Iwata, A., et al., "ATM Connection and Traffic Management Schemes for Multimedia Internetworking," *Communications of the ACM*, Vol. 38, No. 2, Feb. 1995, pp. 72–89. A team from an NEC research lab plays with LAN-based multimedia workstations and NEC's XATOM ATM switch. They find that routers, TCP/IP, and predictive congestion control are inappropriate to the level of media synchronization required. Advocate QoS in three classes: best effort, guaranteed burst, and guaranteed stream.

Jain, R., "FDDI: Current Issues and Future Plans," *IEEE Communications Magazine*, Sept. 1993, pp. 98–105. An inadvertent history, since 1979, of how *not* to implement a LAN standard. FDDI today is neither "fiber," "distributed," "data," nor an "interface." Quarrels and technological progress have clouded the picture on media, connectors, typology, and software versions (I, II) without the base standard ever gaining mass support.

Jajszczyk, A., and H. T. Mouftah, "Photonic Fast Packet Switching," *IEEE Communications Magazine*, Feb. 1993, pp. 58–65. Optical media offers bandwidth to 30 THz; electronic nodes cannot be employed in the transmission path at these speeds. Paper surveys a number of hybrid (electronic cum optical) fast packet switches. Most are passive star couplers combining either fixed wave or tunable transmitter/receivers with electronic control and buffering. Currently, the biggest technical problem is the lack of an optical equivalent to an electronic buffer.

Johnson, J. T., "NREN: Turning the Clock Ahead on Tomorrow's Networks," *Data Communications*, Sept. 1992, pp. 43–61. A news article describing the planned U.S. Government's National Research and Educational Network (NREN). NREN includes five gigabit testbeds, which by 1996 promise to be operating at 1.2 Gbps or higher.

Johnson, J. T., "ATM Comes Out of the Ether," *Data Communications*, March 1993, pp. 41–42. Product review of Newbridge's Vivid as of 3/93. Distributed approach consists of four components—ATM hubs, ridge boxes that convert LAN packets to ATM cells and vice versa, distributed router server software, and LAN service units—that connect ridges via Ethernet and TR interfaces and act as combination wiring hubs and bridges for LAN segments.

Johnson, J. T., "A Next-Generation Hub That's An ATM Switch," *Data Communications*, March 1993, pp. 49–50. Brief technical overview of SynOptics' Lattiscell customer premises ATM switch.

Johnson, J. T., "ATM Networking Gear: Welcome to the Real World," *Data Communications*, Oct. 1993, pp. 66–86. Excellent state-of-the-market article (c. 10/93). Contains a number of vendor equipment reference charts organized by five types of ATM products: campus and LAN switches, private network switches, routers, DSUs, and adaptor cards. Points out many problems: ATM adaptor cards working only with a particular vendor's switch, no standard for encapsulation, most switches do not support AAL 1 or 2. Also, several innovations: use of DSUs as access muxes, use of the DXI (data interchange interface) whereby LAN packets are formatted (in one or two modes) so they can be recognized by DSUs and routers.

Johnson, J. T., "Plug-and-Play ATM Multimedia," *Data Communications*, Jan. 1995, pp. 110–112. A product review of First Visual Corporation's ATM multimedia offering, especially its media operating system (MOS). It consists of PC ATM adaptor cards for LAN emulation, First Virtual's ATM switch, multimedia server with MOS, and a RAID disk array. It allows a user to add multimedia to an Ethernet with Netware or TCP/IP environment, with MOS redirecting the workstation multimedia calls to the multimedia server.

Johnson, J. T., "Router Vendor Makes Its ATM Vision a Reality," *Data Communications*, March 1994, pp. 51–52. Product review of Cisco's ATM switch, a 16 × 155-Mbps port switch that costs $32,000 and supports AALs 3/4 and 5. Designed to work with Cisco 7000 routers ("Ciscofusion," which requires a $22,000 ATM interface!) to fast connect a Cisco router network. Claims superior routing, flow, and security control as well as specifiable data rates and priorities.

Johnson, J. T., "Cells and Frames Come Together," *Data Communications*, March 21, 1993, pp. 31–32. Product review of Cascade Corporation's B-STDX 9000 hybrid ATM/FR switch. Packet size and QoS can be configured on a port-by-port basis. Managed by SNMP/Openview, the switch costs $30–300K, depending on number of ports and interface types.

Johnson, J. T., "The ATM Circus Gets a Ringmaster," *Data Communications*, March 21, 1993, pp. 43–47. Details ATM Forum activities (c. 3/93) on several fronts, including UNI, NNI, B-ICI, and DXI. Also modifications of Q.93B for multicast and an addressing scheme built on OSI's NSAP.

Kajiyama, Y., N. Tokura, and K. Kikuchi, "An ATM VP-Based Self-Healing Ring," *IEEE Journal on Selected Areas in Communications*, Vol. 12, No. 1, Jan. 1994, pp. 171–178. Argues that VP-based self-healing ring scheme's recovery (at path layer) is more simple and efficient than

CCITT-blessed SONET-based SHRs (media-level). Perhaps so, but is it relevant if one comes with standard hardware and the other doesn't? The likely outcome is that we will have both operating independently, with unpredictable results.

Katevenis, M., S. Sidiropoulos, and C. Courcoubetis, "Weighed Round-Robin Cell Multiplexing in a General-Purpose ATM Switch Chip," *IEEE Journal on Selected Areas in Communications*, Vol. 9, No. 8, Oct. 1991, pp. 1265–1279. An academic proposal for design of an ATM switch chip using unconventional buffer management and cell-scheduling techniques. Among the interesting points: that with ATM the cell-processing rate is so high (LE 1 microsecond per cell) that there is almost no room for software, all processing must be done in hardware; use of "cut-through," where an incoming cell is forwarded directly out of the switch without buffering; buffer space equals throughput × round-trip delay, which, in numbers amounts to 1 KB per km and 1 GB/second; DRAMS are less than optical fiber in cost (e.g., 1 MB of buffer storage for each km of link length), thus "site configurable;" congestion is good—with adequate buffers it allows one to achieve full network throughput (re: cost comparison of DRAMS versus deployed fiber).

Katz, H., J. Müermann, and J.-M. Salles, "Teleaction Services: Standards and Architectures," *IEEE Communications Magazine*, June 1994, pp. 54–65. A review of the standards and architectures that can be used to implement teleaction services.

Kawamura, R., K. Sato, and I. Tokizawa, "Self-Healing ATM Networks Based on Virtual Path Concept," *IEEE Journal on Selected Areas in Communications*, Vol. 12, No. 1, Jan. 1994, pp. 121–127. Authors propose a distributed, preassigned scheme for VP restoral on ATM networks to achieve rapid recovery on single link or node failures. (Scheme's efficacy is less persuasive on multiple failures.) Also propose a "reversionless" restoration cycle to be achieved by a dynamic VP reconstruction scheme, but conclude that it will require a lot more spare resources.

Keough, L., "Crafting Standards for an Imperfect World," *Data Communications*, Dec. 1994, pp. 95–96. Sees the ATM Forum (179 principal members at $10,000/yr, 427 Forum members with access to technical specs and proceedings) as the wave of the standards-making future and dominating "the development of the most important networking technology of this century." Contrasts their pragmatic bent with previous standards bodies.

Kim, B. G., and P. Wang, "ATM Network: Goals and Challenges," *Communications of the ACM*, Vol. 38, No. 2, Feb. 1995, pp. 39–44, 109. Discusses modes of accommodating IP over ATM, ISO layers vis-à-vis ATM, and, in an "assertion" and "counterpoint" format, the issues of scalability, traffic integration, statistical multiplexing, and network simplicity.

Kleinrock, L., "ISDN—The Path to Broadband Networks," reprinted in W. Stallings, *Advances in Integrated Services Digital Networks (ISDN) and Broadband ISDN*, Washington, DC: IEEE Computer Society Press, 1992, pp. 151–156. A sort of cheerleading piece for B-ISDN paid for by DARPA; particularly good on historical developments and the interaction between innovation and the market.

Koester, D., "Hands-On Experience with Asynchronous Transfer Mode (ATM) Technology," *The Telecommunications Review*, MITRE Corporation, 1994, pp. 49–62. Article describes actual throughput tests involving a FORE ATM ASX-100 switch and Sun workstations connected with fiber and TAXI interfaces in ATM star and Ethernet bus configurations. As measured by FTP (not accurate), the TCP/ATM links maxed out at 12 Mbps, the TCP/Ethernet at 8 Mbps with 1,500-byte packets. The FORE Sun interface card, which runs SAR processing in the workstation, bounded the former, the media the latter. Where workstations had unequal SAR capabilities, the lack of effective congestion control caused unacceptable packet loss.

Kollias, E. D. and G. I. Stassinopoulos, "ATM Performance Evaluation Under Transparencies of a Distributed System Environment (DSE)," *IEEE Journal of Selected Areas in Communications*, Vol. 12, No. 6, Aug. 1994, pp. 1059–1071. A RACE-funded study of ATM network management with an emphasis on distributed system environments. Good on all the bad things that can happen within ATM, a technology that embodies, like the different sides of a coin, flexibility and complexity. ATM is clearly a bridge technology trying to bridge the philosophical divides of predictability and rigid QoS with extreme decentralization and bursty, untidy traffic flows (i.e., the PSTN and SNA versus Ethernet and the Internet).

Korostoff, K., "Can X.25 Vendors Make the Switch to ATM?," *Data Communications*, March 21, 1993, pp. 17–18. Speculation on who will dominate the ATM market—the packet switch, mux, central office, or router makers—and risk cannibalizing their present product base. Says that the packet switch makers have a shot as they are used to packetizing, statistical muxing, and serious congestion control, and cites Telematics' ATOM and BBN's Emerald prototypes.

Kozaki, T., et al., "32 × 32 Shared Buffer Type ATM Switch VLSI's for B-ISDN's," *IEEE Journal on Selected Areas in Communications*, Vol. 9, No. 8, Oct. 1991, pp. 1239–1247. Describes a 32 × 32 shared buffer-type ATM switch design; that is, where each address of the shared buffer can be allotted to any output port as the occasion demands, not permanently to one particular output port. Further, the shared buffer memory is arranged in the form of a bit slice, where each VLSI corresponds to each bit of an 8-bit parallel format byte of an ATM cell.

Krishnan, R., and N. F. Maxemchuk, "Life Beyond Linear Topologies," *IEEE Network*, March 1993, pp. 48–54. Argues the case that the more complex mesh Manhattan street MAN network topology (with its higher node costs) is superior in throughput and availability to the IEEE's 802.6 DQDB standard.

La Porta, T. F., and M. Schwartz, "The MultiStream Protocol: A Highly Flexible High-Speed Transport Protocol," *IEEE Journal on Selected Areas in Communications*, Vol. 11, No. 4, May 1993, pp. 519–530. Paper presents the multistream protocol (MSP), a new transport protocol designed to operate over high-speed networks and support high-speed applications. MSP is specified as a set of functions that are enabled according to the needs of the application and network being used and are capable of parallel implementation.

La Porta, T. F., et al., "B-ISDN: A Technological Discontinuity," *IEEE Communications Magazine*, Oct. 1994, pp. 84–97. Interesting article explores the "discontinuities" offered by BISDN/ATM in switching, operations and management, signaling protocols and architecture, control methods, and network computing. Many of the effects are thorny technical problems, but many are opportunities, such as using ATM as a workstation bus protocol.

Lane, J., "ATM Knits Voice, Data On Any Net," *IEEE Spectrum*, Feb. 1994, pp. 42–45. Adequate overview of ATM and well-written. Introduces several apt phrases such as ATM being "equitably inefficient" for voice and data and that ATM is "label (address) multiplexing" as opposed to STM's "position multiplexing."

Layland, R., "Unfinished Business: A Theory of Evolution for ATM Technology," *Data Communications*, March 1993, pp. 89–92. A data communications-oriented and sobering view of the ATM standard, identifying a number of areas that compromise efficiency, or, by the incompleteness of the standard, will delay mixed-vendor implementations.

Layland, R., "ATM Without Routers? It Won't Happen," *Data Communications*, Jan. 1995, pp. 31–32. Says that ATM will not be a success until it can handle current broadcast LANs (LAN emulation). And this involves mapping MAC addresses to ATM and providing a way to handle broadcast and multicast traffic. He predicts that routers will be built into these products to keep extraneous broadcasts and multicasts off the WAN.

Layland, R., "Can the ATM Forum Avoid OSI-fication?," *Data Communications*, Dec. 1995, pp. 87–88. Layland criticizes the ATM Forum for (1) an AAL 3/4 spec that is unrelated to how computers work, (2) an AAL 1 spec that wastes bandwidth, and (3) a QoS spec that is too complex to implement. He suggests, to avoid another OSI-like debacle, the Forum should follow IETF practice and require a real product implementation prior to proposing a standard.

Le Boudec, J.-Y., E. Port, and H. L. Truong, "Flight of the FALCON," *IEEE Communications Magazine*, Feb. 1993, pp. 50–56. Contends that the early ATM implementations will be in private networks and that LAN emulation will be the paramount application support mechanism. Predicts older LANs will stay around via reconfiguration (i.e., one station, hub-attached Ethernets). Also sees 100 Mbps on STP/UTP LANs and ATM interfaces at 25 Mbps on UTP. Article describes FALCON, an IBM-Zurich implementation of the private, LAN emulation ATM idea.

Lee, B. G., M. Kang, and J. Lee, *Broadband Telecommunications Technology*, Norwood, MA: Artech House, 1993. Book began as a series of graduate school lectures at Seoul National University in 1987. It has a technologically positivist, pedagogical tone as well as excellent schematic diagrams. It takes broadband networks from the bottom up, with chapters on fiber optics, SDH/SONET, BISDN/ATM (especially nice on ATM switch designs), and high-speed data and video.

Leslie, I. M., D. R. McAuley, and D. L. Tennenhouse, "ATM Everywhere?," *IEEE Network*, March 1993, pp. 40–46. A largely abstract discussion whose thesis is that ATM offers the opportunity to unify communications across a number of dimensions: upper-level traffic types, operating environments, and administrative domains.

Lippis, N., "Hot Switches, Hard Choices," *Data Communications*, Jan. 1995, pp. 25–26. Details the many startups in the switched Ethernet-to-ATM Datacomm market: Agile Networks, Lightstream Corporation, Whitetree Network Technology, Xedia Corporation, and Xylan Corporation.

Malley, D. P., and O. K. Tonguz, "Fiber in the Loop: Where and When is it Feasible?," *IEEE Journal on Selected Areas in Communications*, Vol. 10, No. 9, Dec. 1992, pp. 1523–1544. A most interesting economic analysis of the cost benefits of fiber to the home/business. Although written in 1992, several of the key assumptions have already been overtaken by events, namely that the BISDN delivery mechanism and media would be integrated with phone service and that RBOCs would not join with the cable/home entertainment industry. These events destroy their analysis, and the prevalence of satellite dishes in rural areas probably undercuts their conclusions for that market as well. Their observations on the cost and difficulty of providing emerging power to FTTH shows why home distribution will continue to be centrally powered copper (with emergency backup) for narrow band services and commercial powered cable for BISDN services.

Mandeville, R., "The ATM Stress Test," *Data Communications*, March 1995, pp. 68–82. The author tested seven ATM switches (configured with four 155-Mbps ports and multimode fiber interfaces) for latency, jitter, and performance under "stress," defined as switching delay-sensitive and bursty traffic across one port. FORE Systems' Forerunner ASX-200 is the best performer.

Mandeville, R., and J. T. Johnson, "Forget the Forklift," *Data Communications*, Sept. 1996, pp. 121–134. A European Network Lab (ENL) and Netcom test of LANE. Test scenarios required communications across two Ethernet switches and two ATM switches. Fifteen vendors were solicited with regrets from Agile, DEC, Madge, Netedge, Scorpio (now U.S. Robotics), UB, Whitetree, Whittaker, and Xylan. Six showed: Bay, Cisco, FORE, IBM, Newbridge, and 3Com. Bay and Newbridge sported nonstandard LANE software; Newbridge offered prestandard MPOA. Cisco's Catalyst 500 and Lightstream 1010 and 3Com's Cellplex 7000 emerged as clearly the top performers.

Marks, D. R., "A Forum's Work Is Never Done," *Data Communications*, April 1995, p. 130. Observes that the ATM Forum, though laying the framework for interoperability, has a long way yet to go—APIs, LUNI, PNNI, and flow control.

Marshall, G., "Classical IP Over ATM, A Status Report," *Data Communications*, Dec. 1995, pp. 103–110. One of the best articles comparing IETF 1577 versus LANE. Provides a step-by-step description of how 1577 works. Says that for all-IP LANs, 1577 is the best now because of longer packet lengths and lower protocol overhead. LANE, which also handles

IP, is a more versatile solution and preferable with LAN switches with ATM uplinks. 1577 has some serious problems when one must pass traffic between different logical IP subnets (LIS). Typically, two routers and two VCs are required. (Some routers, called "one-armed routers," can be configured to route packets in and out of the same ATM interface.) Two groups, IETF's IP Over Large Clouds and ATM's MPOA, are working on solutions. One likely outcome, says the author, is that there will be a "cut-though" routing scheme in which routing databases will be accessible for VC call setup.

Matsunaga, H., and H. Uematsu, "A 1.5 Gbps 8 x 8 Cross-Connect Switch Using a Time Reservation Algorithm," *IEEE Journal on Selected Areas in Communications*, Vol. 9, No. 8, Oct. 1991, pp. 1308–1317. Authors propose a design for an ATM cross-connect transit node, switching via VPIs. Classify ATM architectures by position of allocated memories (buffers): input, central, and output. Argue that because central and output buffers must operate equal to port speed times number of ports, input buffers, which operate at nearly port speed, are a natural for high-speed cross connects. They then describe a prototype SD-SW employing a Batcher-Banyan switching network and a time reservation algorithm. Claim their measured maximum port speed to be 1.55 Gbps.

McKinney, R. S., and T. H. Gordon, "ATM for Narrowband Services," *IEEE Communications Magazine*, April 1994, pp. 64–62. The focus is on ATM for narrowband services, not defined but 2 Mbps. Provides an interesting technology management classification: cost-based replacement strategy (stored program control, digital switching) versus revenue-based overlay strategy (cellular, frame relay). Sees ATM "fit" in both deployment strategies, though dependent on compression and lower local access costs, and that demand video will be on the front end (revenue-based) and POTS on the tail end (cost-based, via law of large numbers, mux efficiency). Contains impenetrable discussion of frame relay to ATM adaptation standards.

McDonald, J. C., "Public Network Integrity—Avoiding a Crisis in Trust," *IEEE Journal on Selected Areas in Communications*, Vol. 12, No. 1, Jan. 1994, pp. 5–12. McDonald argues that the forces of technology, regulation, and customer demand have conspired to push the network away from acceptable levels of integrity and that we must think in terms of multiple failures, especially involving software. Public networks should use a societal impact-weighed availability measure, and he suggests the "user lost erlang." He recommends multiple vendor, heterogeneous networks to achieve, covertly, *n*-way programming.

McDysan, D., and D. Spohn, *ATM: Theory and Application*, New York, NY: McGraw-Hill, 1995. A big (595 pages) volume with sweeping breadth. Intended as a one- or two-semester data communications textbook, it covers the OSI model, TDM, SONET, ISDN, network design, kinds of switching and multiplexing networks, X.25, TCP/IP, B-ISDN/ATM standards, and more, as well as 13 chapters on ATM technology. A unique feature is its Chapter 10 on real ATM products. The two authors, from MCI and Southwest Network Services, argue for versatile, high-speed *switched* data networks to achieve a service ubiquity comparable to today's voice networks.

Mier, E. E., "Product Lineup, ATM Switches, Building A New Order," *Communications Week*, June 12, 1995, pp. 67–84. Excellent review of the ATM market as of 6/95. Cites more than 40 models of ATM switches from more than 30 vendors and more than 100 models of workstation ATM adaptors by 16 vendors. As for market tendencies, notes the rapid growth of standards-based SVC, congestion control, LANE 1.0, and the popularity of OC-3 interface, with lukewarm support for the 25-Mbps option and the fading of the 100-Mbps TAXI. Points out, too, that workstation ATM compatibility—especially bus speeds—is a mixed bag with ISA at 30 Mbps, EISA at 250 Mbps, and PCI at 1 Gbps.

Mier, E. E., "ATM To The Desktop: Don't Forget The Obstacles Present," *Communications Week*, June 12, 1995, p. 35. In a discussion of a 25-Mbps ATM versus 100-Mbps FDDI performance comparison, Mier points out that the context of comparison is 25 Mbps maximum in either direction at a time of dedicated access versus 100 Mbps maximum of shared access.

Miller, A., "From Here to ATM," *IEEE Spectrum*, June 1994, pp. 20–24. A highly personal survey of LAN and WAN technologies for data from the 1970s–1990s that ignores SNA, DECnet, TDM, and so forth, but with close attention to wiring.

Minoli, D., *Broadband Network Analysis and Design*, Norwood, MA: Artech House, 1993. A textbook (the author is from Bellcore and teaches at NYU) surveying analytical design methods (financial optimization, queuing and teletraffic techniques, and modeling, performance analysis, "relaxation theory," QoS management) for network design, with the caveat that design mistakes cost more in the broadband environment.

Morency, J., "Legacy LANs Live On Due to Advent of ATM Emulation Service," *Network World*, Feb. 20, 1995, p. 43. Morency explains how the ATM Forum's LAN emulation (called L-UNI or "loony") is being structured, its benefits, and the important caveat that version 1.0 of the specification, due in early 1995, will not provide virtual LAN support for mixed LAN technologies.

Mouchtaris, P. N., "Traffic Analysis for a Custom Video Architecture," *IEEE Journal on Selected Areas in Communications*, Vol. 10, No. 9, Dec. 1992, pp. 1421–1426. Models an ATM-based, customized video news service delivery system employing both central and local storage facilities.

Naegle, J. H., et al., "Building Networks for the Wide and Local Areas Using Asynchronous Transfer Mode Switches and Synchronous Optical Network Technology," *IEEE Journal on Selected Areas in Communications*, Vol. 13, No. 4, May 1995, pp. 662–672. Describes an ATM testbed at Sandia Labs locations in Albuquerque and Livermore. Used a variety of ATM switches, AAL 5, IP, SONET OC-3 @ 155 Mbps, PVCs, and nonstandard TAXI @ 140 Mbps for data transfer. Achieved 71-Mbps memory to memory between workstations. Had problems with the lack of an ATM analyzer, coupling alignment with MM and SM fiber-optic cable, STM1/SDH and OC-3 SONET differences, PVCs, early NICs without on-board processors, IP encapsulation pre-RFC 1483, AAL settings on NICs, and lack of flow control overloading an output buffer, causing TCP retransmissions and freezing a video application.

Neufeld, G. W., et al., "Parallel Host Interface for an ATM Network," *IEEE Network*, July 1993, pp. 24–34. Describes the design of a parallel protocol processing system for a host system connected to an ATM network. Uses standard components (UNIX, VME bus, Motorola 88100 RISC processor) proposing to do lower level protocol processing in hardware, upper level (ISO, TCP/IP) in software. Simulations indicate parallelism possible with both protocols, though easiest with ISO.

Newman, P., "Capitalists, Socialists, and ATM Switching," *Data Communications*, Dec. 1994, pp. 126. An amusing piece contrasting capitalistic circuit switching (telecom) and socialist packet switching (Datacomm) and how they've found peace and unity in the ATM technology.

Newman, P., "Traffic Management for ATM Local Area Networks," *IEEE Communications Magazine*, Aug. 1994, pp. 44–50. First divides ATM service into "capitalist" (guaranteed or constant bit rate) versus "socialist" (variable, "best effort," or available) services. Sees the technical action is in the latter, especially supporting LAN traffic. Maintains that the majority of LAN/data users will not know what their traffic will be, thus, in switch design, tradeoffs between efficiency versus isolation. Sees three kinds of loss-based schemes: high-layer protocol, loss, or delay. On feedback mechanisms, credit-based versus rate-based. Proposes a BECN-based scheme as it takes effect faster and is network-based, but admits either feedback based approach is workable.

Newman, R., "You Don't Have to Start from Scratch," *Internetwork*, Jan. 1996, pp. 40–44. The writer, from Bay Networks, argues the case for ATM scalability in the LAN environment and sees ATM as solving "the general aggregation problems." Says that 80/20 rule (local:distant) violations may be more frequent (web surfing and videoteleconferencing) and says that most of today's ATM switch designers assume 3:1 or 4:1 between UNI and NNI traffic, but most switch ports are designed to be interchangeable.

Nishio, M., et al., "A New Architecture of Photonic ATM Switches," *IEEE Communications Magazine*, April 1993, pp. 62–68. Article proposes a new architecture of photonic ATM switch for the purpose of realizing ATM operation at throughputs greater than 1 Tbps. The proposed architecture is based on a vertical to surface transmission electrophotonic device (VSTEPS), which can be used as both an optical buffer memory and an optical self-routing circuit.

Nolle, T., "Broadband," *Communications Week*, March 13, 1995, pp. S4–S6. Author focuses on the economics and availability of broadband services. Claims that the cost of T3 has fallen 25% in two years, and the T1/T3 crossover point is now 9 Mbps.

Onvural, R. O., *Asynchronous Transfer Mode Networks, Performance Issues*, Norwood, MA: Artech House, 1994. Onvural's volume addresses various ATM performance issues with a sort of manifesto: "The experience we have gained from circuit-switched networks for about a century and packet-switch networks for several decades has not been very helpful in addressing the congestion problem in networks with large propagation delay-bandwidth product links" (p. xiv). Requires background in stochastic process, especially queuing theory. Contains a valuable discussion of "lightweight" protocols.

Paone, J., "Nobody's Banging Down the Doors," *Internetwork*, Jan. 1996, pp. 21–23. Author, from OnStream Networks, observes that though ATM may rule the future, due to no internetworking between carriers, emerging standards, and standards under development, frame relay rules today's WANs.

Partridge, C., *Gigabit Networking*, New York, NY: Addison-Wesley, 1994. The book, which should be called *Problems Facing Gigabit Networking*, focuses on the outstanding research issues from an academic, data networking, TCP/IP, and UNIX computer performance perspective.

Patel, S. H., "Congestion Control and Performance Issues in Asynchronous Transfer Mode (ATM) Networks, ¡*The Telecommunications Review*, MITRE Corporation, 1994, pp. 109–118. Basically a discussion of some well-known ATM problems at a high level such as statistical multiplexing (good definition), fairness, call admission, congestion control, and source policing. Contains some screamers relative to the networks of the 1960s–1990s.

Pathak, G., et al., "Integrated Network and Service Management for the NCIH," *IEEE Network*, Nov./Dec. 1994, pp. 56–63. Describes NCIH network management tools: the TelOps Network Integrated Control System (TONICS) and the Broadband Service Management System (BSMS). Net management was bedeviled by having to switch PVCs as SVCs, incomplete CMIP messages on the FETEX, no usage measurement, beta versions of everything, and a constantly evolving hardware and software environment.

Patrick, P., "Transaction-Oriented Applications via National ISDN," *IEEE Communications Magazine*, June 1994, pp. 44–48. Describes the benefits of national ISDN use for dial-up transaction processing applications, especially the use of TP3 POS over X.25, lowering transaction times from 20–30 seconds to 7 seconds.

Pattavina, A., "Nonblocking Architectures for ATM Switching," *IEEE Communications Magazine*, Vol. 31, No. 2, Feb. 1993, pp. 38–48. The author makes four observations: (1) pure input queuing is easier to implement than pure output or shared queuing in a nonblocking ATM switch architecture; (2) output or shared queuing architectures provide optimal delay-throughput performance, whereas the pure input queuing architecture suffers severe throughput degradation due to the HOL blocking phenomenon; (3) mixing two different queuing strategies aims at coupling the advantages peculiar to the single strategies adopted alone; and (4) that the target of providing a certain loss and delay performance always can be accomplished for each of these architectures by selecting a proper level of the offered load.

Perloff, M., and K. Reiss, "Improvements to TCP Performance in High-Speed ATM Networks," *Communications of the ACM*, Vol. 38, No. 2, Feb.

1995, pp. 90–100. Simulations on TCP/IP performance find that fair link utilizations to 85% are possible with large (4K cell) buffers and additional discard-based congestion control. But TCP/IP performance can be horrible and tuning is essential.

Perros, H., (ed.), *High-Speed Communications Networks*, New York, NY: Plenum Press, 1992. A collection of 19 papers on the topics of local ATM, congestion control, standards, routing, transport protocols, traffic measurements, and telecommunications software engineering, presented at TriComm '92, the fifth in a series of Research Triangle conferences on computer communications, held Feb. 27–28 in Raleigh, NC.

Perry, T. S., (ed.), "The U.S. HDTV Standard—The Grand Alliance," *IEEE Spectrum*, April 1995, pp. 36–45. Describes the "Grand Alliance's" emerging HDTV standards. Audio is at about 384 Kbps. Video uses several formats, 1,920 × 1,080 or 1,280 × 720 and 24, 30, or 60 frames per second for around 20 Mbps with compression. Most interesting, they use the MPEG-2 transport system, a cell-based 188 (4/184) system.

Petrosky, M., *Asynchronous Transfer Mode: The Burton Group Report*, Salt Lake City, UT: The Burton Group, Vol. 1, Sept. 1994. A very thoughtful introduction to the ATM technology and market place (c. 3Q 1994). Author concluded, per the very incomplete state of the standards at that time—that ATM was not ready for Prime Time. Further, in assembling all the needed pieces to make ATM work as hyped, it was less the case that "your mileage will vary," but that some legacy software or hardware widget in the data path will assure that measured performance will be considerably below expectations.

Petrosky, M., *ATM Market Maturing: The Burton Group Report*, Salt Lake City, UT: The Burton Group, Vol. 1, May 1996. Focuses on the LAN/ATM market and provides detailed descriptions of 3Com, Bay, Cabletron, Cisco, DEC, FORE, IBM, Madge, and Newbridge product developments. At a higher level, makes a number of telling points about the relationships between standards, technology, and the ATM industry: (1) even with the Anchorage Accord in 1996, 1999 is her target date for a mature ATM technology; (2) with mergers, buyouts, and licensing deals, the industry is extremely unstable; (3) at least annual hardware and software upgrades for ATM products for as far as the eye can see; (4) at least into 1997, single vendor shops to ensure that ATM works as advertised; (5) when immediate functional needs clash with long-developing standards, prestandard solutions are guaranteed; (6) the more complex standards (UNI, PNNI) take two years to go from specification to fielding.

Petrosky, M., and J. Lewis, "3Com ATM Roll-Out," *The Burton Group*, March 6, 1995, unpaged. Product review of 3Com's Cellplex 7000 ATM backbone switch, the LinkSwitch 2700, and Cellplex 7200 combined Ethernet and ATM workgroup switches. Says that 3Com was late to market on LAN/ATM switching, but in doing so fielded more standards-conforming products and a more end-to-end approach.

Petrosky, M., and J. Lewis, "FORE Announces Switches, Updates Architecture," *The Burton Group*, Oct. 18, 1995. A thoughtful review of the FORE product line, and, more widely, the FORE business philosophy. Items: (1) the ADC Kentrox cross-licensing deal; (2) FORE's entry into Ethernet switches via applied network technology; (3) their 622-Mbps speed king backbone switch, the ASX-1000; (4) "prestandard" software versus the costs of maintaining proprietary and standards-based versions; (5) plans to integrate LAN/WAN QoS and routing; (6) more merger mania, including Rainbow Bridge Communications for distributed routing expertise; (7) "value-added" technology versus interoperable technology; and (8) technical elegance versus standards.

Petrosky, M., "The Anchorage Accord Promises to Stabilize ATM," *Network World*, May 6, 1996, p. 39. In this editorial piece, she recounts the ATM Forum's mid-April 1996, "Anchorage Accord." Says since 1991 the Forum has approved nearly 50 specifications, 20 more are due by the end of the summer of 1996, and dozens more are in progress—and the market can't absorb it. Accord calls for backward compatibility and identifies two classes of specifications—"foundation" and "expanded features." Foundations are "core LAN and WAN ATM specifications, such as UNI signaling, the private network-node interface (P-NNI), the broadband intercarrier interface (B-ICI), traffic management, and key physical layer interfaces. The expanded features specifications will include LAN emulation, circuit emulation, audiovisual multimedia service, and voice over ATM." But there are a number of problems: (1) the "foundation won't be fully defined until early next year, and compliant products won't be available until 6 to 12 months later;" (2) "many ATM specifications are interdependent, and a change in one requires a change in others;" (3) "many switches will require a hardware upgrade in order to support the recently approved rate-based traffic management scheme;" and (4) "it won't guarantee that vendors will support standards in a timely fashion. Although UNI 3.1 was finalized in Sept. 1994 and UNI 3.0 a year before that, both Cisco Systems, Inc. and FORE Systems, Inc. only recently announced support for them."

Phillips, B., "ATM: From Sea to Shining Sea" *Data Communications*, Oct. 1993, pp. 127–128. New service offering from Sprint, a 45-Mbps ATM PVC network with 307 POPs. POPs are connected to a ring of six TRW 2010C Broadband Access Switches. Users can pay flat rate or usage-based; rates vary by QoS. Estimates are that prices are 30–50% less than a leased DS-3 network. Routers are connected to the backbone by Digital Link Corp. ATM DSUs over a HSSI port.

Phillips, B., "Hot Products—Data Transmission and Switching: Let the ATM Evolution Begin" *Data Communications*, Jan. 1994, pp. 43–45. Product review of StrataCom's BPX ATM edge switch. Versatile—will connect with packet, cell, and circuit interfaces, public and private networks, as well as StrataCom's proprietary IPX frame relay equipment. Employs a 9.6-Gbps crosspoint switching matrix and claims switch can handle 20 Mcells/s. Offers CCITT I.371 "open loop" congestion control (devices that feed the net have no knowledge of the status of the net) as wells as proprietary "Foresight" "closed loop" (communicating switches) scheme. Extends ATM's four standard QoS classes to 32 ("Opticlass"), allowing each connection to have its own QoS. BPX's "Autoroute" does automatic connection management and automatic reconfigure by specifying source and destination addresses. High-end BPX costs $500,000, used by AT&T's Interspan service. Offers "Reliaburst" which allows ABR within CBR, "Fastar" automatic rerouting, and SNMP-based network management.

Phillips, B., "ATM Access Over T1: No Longer a Pipe Dream" *Data Communications*, Feb. 1994, pp. 103–104. Product announcement of AT&T's Interspan ATM service, offering 300 locations by mid-1994 and speeds as low as T1 (using header error control cell mapping method à la ITU Recommendation G.804 for about 10% overhead). Service offers PVCs with AAL 1 and 5 and uses Stratacom's BPX/IPX at the POPs and AT&T Network systems GCNS-2000 ATM switches at the major nodes. Potentially, will compete against AT&T's Accunet fractional T1 service.

Phillips, B., "ATM Meets TDM in Hybrid Switch" *Data Communications*, Nov. 1993, pp. 45–46. Product announcement of an interesting hybrid switch, the 6950 Softcell ATM Networking Node, from Motorola Codex. Similar to IBM's TNN, which handles both FR and ATM and uses variable-length cells, the 6950 combines TDM and ATM and switch-adjusted, port-variable, variable-length cells. The idea is to provide easier migration from TDM-based WANs. $60,000–$140,000, depending on size and features.

Rarig, H., "ISDN Signal Distribution Network," *IEEE Communications Magazine*, June 1994, pp. 34–38. Describes an ISDN application whereby an ISDN network is used as a flexible distribution medium for directing real-time signals over a WAN to a staged transport server. Could be used for remote recording session or video distribution.

Romanow, A., and S. Floyd, "Dynamic of TCP Traffic over ATM Networks," *IEEE Journal on Selected Areas in Communications*, Vol. 13, No. 4, May 1995, pp. 633–641. Authors address the problem of TCP/IP "fragmentation" with AAL 5 over ATM. "Fragmentation" is when a cell is dropped by the switch, retransmission occurs, and throughput plummets. After simulations, they suggest a number of palliatives: larger port buffers and smaller packet sizes; partial packet discard (PPD), and better early packet discard (EPD). Further, they suggest modifications to TCP to improve its performance over ATM; they refer to mechanisms such as "drop-tail" packet dropping (with an output buffer overflow, the packets arriving at the output buffer are dropped) and random early detection (RED) gateways that are designed to maintain low average queue sizes. Also point out that ATM AAL 5 "best effort" (ABR) error rates are currently undefined.

Rooholamini, R., V. Cherkassky, and M. Garver, "Finding the Right ATM Switch for the Market," *Computer*, April 1994, pp. 16–28. An overview of the ATM standard, switching fabrics, current products, but with little critical comment or analysis. Good on the basics of switch design trade-offs. They imply that current switches would have a high cost of ownership, but again not specific. Believe the optical scaling design would be a hybrid space/time division switch, but still rather shy on why and what.

Ross, T. L., "ATM APIs: The Missing Links," *Data Communications*, Sept. 1995, pp. 119–124. Good tutorial on the role of APIs. Today, without ATM APIs, the user has two choices. Hide ATM by using the LAN emulation user-network interface (LUNI) or the IETF's classical IP over ATM. Wait for O/S or applications vendors to extend present APIs like UNIX sockets or winsock. The Forum's Service Aspects and Applications Group is working to define a native ATM API. The SAA proposal defines three service access points (SAPs): Type 1 for LANE, Type 2 for classical IP, and Type 3 for native ATM.

Roth, B., et al., "An Evolving Methodology for ATM/SONET Network Modeling and System Planning," *The Telecommunications Review*, MITRE Corporation, 1994, pp. 91–100. Authors explain that traditional communications modeling techniques using discrete event simulation are inappropriate for high-speed ATM environments where current models may need hours to simulate seconds of operation. Advocate data stream modeling methods that focus on changes in the aggregate rate at which packets enter and exit the resource. They then compare input and output cell rate wave forms for quality of service, network flow, and capacity approximation.

Ruiu, D., "ATM at Your Service?," *Data Communications*, Nov. 1993, pp. 85–88. Ruiu points out that unless ATM gets service-level definitions on details like billing and management, ATM carrier services will be either incomplete or proprietary—and uses ISDN as a negative example. Historically, he describes an AAL 0 or null as one where CPE assumes all AAL functions, then maps AAL 1 to fractional or full T1/T3, still-to-be defined AAL 2 to compressed data, AAL 3 to FR, AAL 4 to SMDS (later merged with AAL 3), and AAL to 5 to a few featured, simple to implement connectionless data service (like TCP/IP).

Ruiu, D., "Testing ATM Systems," *IEEE Spectrum*, June 1994, pp. 25–27. Testing ATM will require elegant equipment because of ATM's speed, muxed approach to traffic, and Q.2931's signaling power and complexity. HP to the rescue!

Sachs, M. W., A. Leff, and D. Sevigny, "LAN and I/O Convergence: A Survey of the Issues," *Computer*, Dec. 1994, pp. 24–32. LAN and I/O channel distinctions, previously due to distance and computational model (master/slave, peer/peer), are becoming fewer—perhaps on the way to none.

Saito, H., *Teletraffic Technologies in ATM Networks*, Norwood, MA: Artech House, 1994. The book, from an engineer with NTT Laboratories, explains the ìteletraffic technologies" employed by ATM (the 1992 CCITT version). These include usage parameter control (UPC) and connection admission control (CAC) relative to "preventive control" (policing), and cell delay variation (CDV) relative to resource management. The book also contains a brief overview of ATM technology and statistical multiplexing and concludes with a discussion of "dimensioning" (VP bandwidth to meet a connection-loss probability objective).

Salamone, S., "A Leading-Edge Router for LAN Interconnect," *Data Communications*, Feb. 1994, pp. 39–40. Product review of Netedge's ATM Connect edge router, connecting Ethernets, TRs, FDDI LANs with Appletalk, XNS, TCP/IP, IPX, and handling BGP, EGP, OSPF, RIP, and RTMP, and connecting ATM T-3 UNI with an integral CSU/DSU for $42,500. Employs three i960 RISC chips in an asymmetrical hierarchy.

Sathaye, S. S., *ATM Forum Traffic Management Specification Version 4.0 (Draft)*, Warrendale, PA: FORE Systems, July 25, 1995. A very readable document with considerable helpful introductory material on perhaps the thorniest technical problem with the ATM technology—traffic management. Also an indication intellectually how far the road goes—and needs to go—from the ITU-T conception to ATM that does useful work. Gives one a sense of how really complex the traffic management software will be and, with the number of "network specific" qualifiers, how elusive heterogeneous vendor interoperability may prove. Section 5 (pp. 39–61) discusses the various traffic and congestion control functions in detail.

Sato, K.-I., S. Ohta, and I. Tokizawa, "Broad-Band ATM Network Architecture Based on Virtual Paths," reprinted in W. Stallings, *Advances in Integrated Services Digital Networks (ISDN) and Broadband ISDN*, Washington, DC: IEEE Computer Society Press, 1992, pp. 208–218. See: Aoyama, T., I. Tokizawa, and K.-I. Sato, "Introduction Strategy and Technologies for ATM VP-Based Broadband Networks."

Saunders, S., "A Mix-and-Match Switch for Ethernet and ATM," *Data Communications*, March 1995, pp. 43–44. Product review of the Whitetree Network Technologies WS3000 Workgroup switch, a LAN switch that handles switched Ethernet or 25.6-Mbps ATM on each port for $7,795 for 12 ports.

Saunders, S., "ATM LAN Switch Puts SVCs Into Service," *Data Communications*, April 1994, pp. 41–42. Product review of FORE's LAX-20 ATM LAN switch. It's advantages are that it can do SVCs to FORE ASX-100s and that it supports LAN emulation with Ethernet, TR, and FDDI. It's disadvantages are that its Simple Protocol for ATM Network Signaling (SPANS), SVC, and LAN emulation techniques are proprietary. (On SVCs, it is capable of Q.2931.)

Saunders, S., "Closing the Price Gap Between Routers and Switches," *Data Communications*, Feb. 1995, pp. 49–50. Product review of Netedge's ATM Connect Edge Router, a modular router that switches traffic between TRs and Ethernets, FDDI or ATM LAN nets, as well as serial WAN links.

Saunders, S., "Desktop ATM: It's In The Cards," *Data Communications*, Jan. 1995, pp. 94–96. Describes IBM's currently proprietary ATM LAN emulation product, 25.6-Mbps Turboways ATM adaptors (for ISA workstations) at $405 per unit, $30 less than 16-Mbps token ring cards. They attach to a Turboways 8282 ATM workgroup concentrator allowing twelve 25-Mbps workstations to share a 100-Mbps TAXI on the ATM switch module of its 8260 hub. Cost: $705 per port versus $1,800 per port for the least expensive 16-Mbps TR switch.

Saunders, S., "Making Virtual LANs a Virtual Snap," *Data Communications*, Jan. 1995, pp. 72–74. A product review of Agile's ATMizer 125 Ethernet/ATM switch that automatically assigns end stations to virtual workgroups based on network protocol and subnet address information. Disadvantages include (1) routers are needed to transport traffic to different virtual workgroups, (2) FDDI is not supported, and (3) traffic analysis is difficult or expensive as each port must be tested separately.

Saunders, S., "Making the ATM Connection for Ethernet Workgroups," *Data Communications*, March 1994, pp. 47–48. Product review of Synoptics' Ethercell switch, which connects to workstations with UTP/RJ45, and to the Lattiscell Model 10114 ATM switch with multimode fiber at 155 Mbps using AAL5. The intent is to connect virtual LANs at high speed and low cost, though high latency is a problem.

Saunders, S., "ATM Forum Ponders Congestion Control Options," *Data Communications*, March 1994, pp. 55–60. Clearly written article on the state of ATM congestion control as of 3/94. Points out that with UNI 3.0, CBR uses a reserved, fixed-rate connection with one leaky bucket and VBR uses the same with two leaky buckets, one for sustained, one for burst. These are commonly augmented by use of a cell loss priority bit and/or priority queues or oversized buffers. More challenging is VBR and the debate over FECN, BECN, and credit schemes. StrataCom's proprietary VBR uses FECN ("Foresight") and claims 90–95 capacity. Good quote by Keith McCloghrie of Hughes: "A very small cell loss results in a huge frame loss, which translates to huge delays."

Saunders, S., "Choosing High-Speed LANs," *Data Communications*, Sept. 21, 1993, pp. 58–70. Valuable article on the state of the LAN market as of 9/93. Points to several major tendencies: LAN adaptor costs as king, the end of shared-media LANs (and their replacement by LAN switches), the role of high labor costs ($350 for UTP install, $1,800 for multimode fiber install, $1,100 to install a new adaptor card & driver), and the role of IEEE standards groups (standards proliferation via self-serving vendors, lengthy time to market).

Schmidt, D. C., "Transport System Architecture Services for High-Performance Communications Systems," *IEEE Journal on Selected Areas in Communications*, May 1993, Vol. 11, No. 4, pp. 489–506. Schmidt notes the mismatch between the performance of the network and transport

infrastructures and points out that "transport system overhead is not de-
creasing as rapidly as the network channel speed is increasing. This results from improperly layered transport system architectures…and is also
exacerbated by the widespread use of operating systems such as UNIX,
which are not well-suited for asynchronous, interrupted-driven network
communications." p. 490) He later goes on to describe a number of transport services that attempt to creatively incorporate operating resources
(CPUs, virtual memory, I/O devices) in order to achieve high performance
with distributed applications.

Schnepf, J. A., et al., "Building Future Medical Education Environments
Over ATM Networks," *Communications of the ACM*, Vol. 38, No. 2, Feb.
1995, pp. 54–69. A prototype medical distance learning application using
ATM's AA1 1 and 3/4. Long on human interface and technical problems
and extremely expensive.

Schroeder, M. D., et al., "Autonet: A High Speed, Self-Configuring Local
Area Network Using Point-to-Point Links," *IEEE Journal on Selected
Areas in Communications*, Vol. 9, No. 8, Oct. 1991, pp. 1318–1335. Describes a prototype 100-Kbps high-speed LAN built for DEC's Systems Research Center in Palo Alto, a sort of Ethernet follow-on built around active
crossbar switches. In many cases, such as obligatory network-wide reconfiguration and monitoring link-dependent datagrams, they have relearned
old lessons, or lessons long-ago discovered by others. Is it a case of intellectual hysteresis (to come late, be behind)?

Scott, H., "Teleaction Services: An Overview," *IEEE Communications
Magazine*, June 1994, pp. 50–53. A pitch for "teleaction" services (TP,
alarm and surveillance, business automation, utility resource management, control and command, interactive video support), most of which
would be greatly advanced by N-ISDN.

Shelef, N., "SVC Signalling: Calling All Nodes," *Data Communications*,
June 1995, pp. 123–130. A tutorial article explaining SVC signaling steps
for point-to-point and point-to-multipoint connections. Includes some interesting observations on SVC implementation, especially SVC and port
memory requirements.

Shobatake, Y., et al., "A One-Chip Scalable 8 * 8 ATM Switch LSI Employing shared Buffer Architecture," *IEEE Journal on Selected Areas in
Communications*, Vol. 9, No. 8, Oct. 1991, pp. 1248–1254. Paper describes a one-chip scalable 8 × 8 shared buffer switch LSI for ATM. Seeking to address ATM's main drawback, cell loss, they jointly combine "the

sharing effect," input link speedup, and input slot rotation design tricks to that end.

Smith, J. M., and C.B.S. Traw, "Giving Applications Access to Gbps Networking," *IEEE Network*, July 1993, pp. 44-52. Reports on a number of U. of PA AURORA experiments with IBM RS/6000 workstations on host interface strategies. Presently, there is a mismatch between B-ISDN network speeds and the memory bandwidths of today's workstations, and a number of technical approaches to narrow this gap are effective on test data sets. In general, the objective is to reduce copying and preserve host capacities sufficient to perform applications processing.

Stallings, W., "Asynchronous Transfer Mode (ATM) and Synchronous Optical Network/Synchronous Digital Hierarchy (SONET/SDH)" in *Advances in Integrated Services Digital Networks (ISDN) and Broadband ISDN*, Washington, DC: IEEE Computer Society Press, 1992, pp. 177–181. A nice summary of SONET and ATM and how they fit together.

Stavrakakis, I., "Efficient Modeling of Merging and Splitting Processes in Large Networking Structures," *IEEE Journal on Selected Areas in Communications*, Vol. 9, No. 8, Oct. 1991, pp. 1336-1347. Author provides a number of observations on the effects of packet splitting and merging operations on bursty traffic.

Stevenson, D. S., and J. G. Rosenman, "VISTAnet Gigabit Testbed," *IEEE Journal on Selected Areas in Communications*, Vol. 10, No. 9, Dec. 1992, pp. 1413–1420. Description of the VISTAnet gigabit networking testbed in North Carolina and its networking and oncological objectives. Observes a number of real-world problems in BISDN in HiPPI-to-ATM conversion, network saturation by 3-D graphics at 622 Mbps, and accommodating supercomputer packets sizes up to 1 Mb.

Stevenson, D., et al., "The NC-REN Meets the North Carolina Information Highway," *IEEE Network*, Nov./Dec. 1994, pp. 32–38. Describes the experiences of the North Carolina Research & Education Network (NC-REN) with a distance learning application at 16 campuses. Data used Cisco routers running IGRP and BGP-4 and then IP on SMDS. Video used DS-3 with DV45 coding with a MPEG II (at 6 Mbps) changeover in the future. The base ATM switches were Fujitsu FETEX 150s running CRS and CES. Lots of system integration problems (re: ATM as "adolescent technology") encountered in their pursuit of production-quality distance learning ("virtual proximity").

Suzuki, T., "ATM Adaptation Layer Protocol," *IEEE Communications Magazine*, April 1994, pp. 80–83. A practically impenetrable article on the background of the ATM adaptation layer (AAL) split between the connection-oriented (CO) and connectionless (CL) camps, how each ended up with a unique AAL (types 3 and 4, respectively), how they were merged into a too-complex compromise (type 3/4) and finally de facto replaced (with "minimalist" type 5). Unfortunately, types 3/4 and 5 cannot interoperate without protocol conversion.

Swallow, G., "PNNI: Weaving a Multivendor ATM Network," *Data Communications*, Dec. 1994, pp. 102–110. Describes the upcoming private network-to-network interface (PNNI) that will allow heterogeneous ATM switches to establish SVCs between each other. Expected by 7/95, it is a sophisticated peer to peer source routing protocol that uses UNI 3.1 signaling and automatically creates hierarchies and information about potential communicants. Also described is the interim standard, interim interswitch signally protocol (IISP).

Tarraf, A. A., I. W. Habib, and T. N. Saadawi, "A Novel Neural Network Traffic Enforcement Mechanism for ATM Networks," *IEEE Journal on Selected Areas in Communications*, Vol. 12, No. 6, Aug. 1994, pp. 1088–1096. Describes a prospective neural net (NN) approach to policing ATM variable bit rate traffic. Claims the NN approach more efficient (a single set of parameters track both peak and mean bit rate violations) and that the NN approach has a very short reaction time compared to other policing mechanisms.

Taylor, K. M., "ATM Switching: Freedom of Choice," *Data Communications*, Jan. 1995, pp. 46–48. Review of Lightstream's 2020 ATM switch that interfaces slower speed LAN protocols (Ethernet, FDDI) and point-to-point protocols (HDLC, SDLC) with high-speed WANs (56 Kbps to 2.048 Mbps).

Taylor, K. M., "Starting Small With ATM," *Data Communications*, Jan. 1995, p. 56. Review of ADC Kentrox's ATM multiplexer, which converts voice and video (CBR) and data traffic (VBR) into ATM cells for transmission over T1/E1 links.

Taylor, K. M., "Concentrating on Keeping ATM Prices Down," *Data Communications*, March 21, 1995, pp. 45–46. Product review of ADC Kentrox's ATM Access Concentrator-3. It is a low-priced ($10,000–$60,000) concentrator capable of CBR and VBR and supporting many interfaces, including high-speed serial interface (HSSZ), SMDS data exchange interface (DXI), and ATM DXI. Unique congestion control feature claims to discard cells based on their position within a message, thus killing a message early on.

Taylor, K. M., "Assessing ATM Analyzers," *Data Communications*, Dec. 1995, pp. 93–100. An excellent article overviewing the ATM analyzer market. Currently there are 17 vendors (see tables) offering test gear from $7,000–$150,000. Taylor identifies three kinds of equipment for three different constituencies. At the low end, physical-level testers employed by carriers to install CPE and cable to test WAN interfaces. In the middle, ATM-equipped protocol analyzers to do ISO 1-7 level testing by corporate networkers, especially with legacy protocols. At the high end, research and development analyzers that can generate and test ATM traffic, used by developers for conformance testing.

Taylor, K. M., "Voice Over ATM: A Bad Connection," *Data Communications*, Feb. 1996, pp. 55–59. Reviews the current less-than-desirable state in VTOA (voice and telephony services over ATM). ATM voice via CBR is less efficient than TDM. Methods that use VBR and low bit rate coding with silence suppression are still proprietary and friction between vendors may keep them so for some time. Meanwhile, ITU Study Group 13 is proposing an AAL 6 between a wireless base/mobile switch center that defines low bit rate and silence suppression.

Taylor, K. M., "The Ultimate ATM Mix: More Talk, Less Bandwidth," *Data Communications*, Jan. 1996, pp. 43–44. Product review of Nortel's Magellan Passport ATM switch. Switch uses various tricks to efficiently handle voice: proprietary AAL that defines voice as a VBR-RT; silence suppression; and ADPCM compression at 32, 24, or 16 Kbps, with silence suppression and compression based on load or user QoS. Used by MFS Datanet's WAVE service, savings are in the 14–25% range.

Taylor, K. M., "Top-Flight Video Makes It to ATM," *Data Communications*, Jan. 1996, pp. 52–54. Product review of TRT-Phillips' ATM Node 10,000 ATM switch whose distinguishing feature is that it is the first ATM switch to support MPEG-2 video via CBR, and, to greatly reduce errors, employ ITU I363's RAID-like error correction scheme. (With MPRG-2, a 216-Mbps digital TV signal is compressed to 6 Mbps without sacrificing quality, but coders cost up to $100,000 and decoders run $1,000–$3,000.)

Taylor, S., "Hope Not Hype for Broadband ISDN," *Data Communications*, Oct. 1993, pp. 25–26. Looks at N/B-ISDN from the point of view of the home-bound telecommuter and finds only B-ISDN/ATM technically practicable for today's computing practices, but availability/pricing a question mark.

Taylor, S., "ATM and Frame Relay: Back to the Basics," *Data Communications*, April 1994, pp. 23–24. Attempts to compare FR and ATM on function, costs, and availability. FR is more efficient, ATM more versatile and speedier. ATM's availability is so limited that direct cost comparability is not possible. Concludes that FR is a "valuable transition technology."

Taylor, S., "Broadband Trends to Watch in '95," *Data Communications*, Jan. 1995, pp. 21–22. Points out that given the low current costs for voice, packetized voice may be hard to justify. And that AAL 2, designed for packetized voice and video, is a low priority of the ATM Forum.

Tobagi, F. A., T. Kwok, and F. M. Chiussi, "Architecture, Performance, and Implementation of the Tandem Banyan Fast Packet Switch," *IEEE Journal on Selected Areas in Communications*, Vol. 9, No. 8, Oct. 1991, pp. 1173–1193. Authors propose a tandem switching fabric using multiple copies of the Banyan network in series to (1) provide multiple concurrent paths between inputs and outputs, (2) overcome the problems induced by output conflicts and blocking, and (3) achieve output buffering. Authors argue that packet switching (ATM) is the appropriate technology for high-speed nets because it is superior for bursty data traffic and more flexible than circuit switching in handling the wide variety of data rates and latency requirements resulting from service integration. They present three architectural designs for high-speed packet switches—shared memory, shared medium, and space division—as well as three fabrics of space division switches—crossbar, Banyan, and N * 2 disjoint paths.

Tolly, K., "In Search of ATM LAN Emulation," *Data Communications*, Sept. 1995, pp. 29–30. The Tolly Group (c. 8/95) tried to see whether any ATM switch vendor could run LAN emulation in a commercial Novell Ethernet/token ring server environment. Only FORE, with Netedge ATM routers, and FORE EISA NICs, could demo the capability, and they relied on FORE's proprietary software (viz., LANE protocols, Open Data Link Interface (ODI), LANE version 0.4, and Simple Protocol for ATM Network Signaling (SPANS)).

Tolme, D., and J. Renwick, "HiPPI: Simplicity Yields Success," *IEEE Network*, Jan. 1993, pp. 28–32. A terse, excellent description of the high-performance parallel interface (HiPPI), a simplex point-to-point interface for transferring data at peak data rates of 800 (or 1,600) Mbps using 50 pair (or 2, 50 pair) copper, twisted pair cables up to 25 meters. Basically an ISO 0-2 layer ANSI standard. Also discusses the HiPPI future, which includes HiPPI crossbar switched hub typology LANs and serial-fiber implementations of HiPPI.

Truong, H. L., et al., "LAN Emulation on an ATM Network," *IEEE Communications Magazine*, May 1995, pp. 70–85. A long and excellent article, funded by RACE, describing and evaluating the alternatives relating to LAN emulation (LE) over ATM. It describes the ATM Forum's current LE direction: LE exists as a layer over the MAC that supports separate emulations of Ethernet and token ring. It is part of the UNI spec and uses AAL 5. Basically, for each LAN segment, LEC (clients) communicate with LES (servers) to obtain connection VCs and ATM addresses of other LECs and LES'. They then have session-oriented VCs. If the LES doesn't know the ATM address, it is referred to the BUS (broadcast/unknown server) that issues a LE-ARP (address resolution protocol) supporting either transparent bridging (Ethernet) or source routing (TR). In the case of multicast, the connection request goes to a multicast server with the LES as root and the LECs as leaves. LE doesn't do 802.3 to 802.5 translations. Emulated LANs are connected by routers and bridges (and probably in the future) ATM switches. Configuration for LE is handled by a configuration server (LECS) that first searched the interim local management interface (ILMI) management information base (MIB), then a 411-like facility called an "anycast address."

Tsuboi, T., et al., "Deployment of ATM Subscriber Line Systems," *IEEE Journal on Selected Areas in Communications*, Vol. 10, No. 9, Dec. 1992, pp. 1448–1458. Authors propose an ATM subscriber line terminal (SLT) consisting of service-dedicated virtual paths (VPs) and call-by-call virtual channels (VCs). The presumed applications are database retrieval and video distribution; the user interface issue is not addressed.

Urushidani, S., "Rerouting Network: A High-Performance Self-Routing Switch for B-ISDN," *IEEE Journal on Selected Areas in Communications*, Vol. 9, No. 8, Oct. 1991, pp. 1194–1204. Paper describes a high-performance, self-routing switch where performance is enhanced by applying a rerouting algorithm to the particular multistage interconnection algorithm, which embeds plural Banyan networks within the switch.

Van Landegem, T., P. Vankwikelberge, and H. Vanderstraeten, "A Self-Healing ATM Network Based on Multilink Principles," *IEEE Journal on*

Selected Areas in Communications, Vol. 12, No. 1, Jan. 1994, pp. 139–148. Survey of current "self-healing" methods within ATM (automatic protection switching, protection switching at VP level, self-healing rings, flooding algorithms) and advocates the "multilink" concept. With the multilink concept, cells of connections are distributed over several physical links. If failure occurs, the cells are redistributed over the surviving physical links. Authors claim that multilink concept is, via stat-mux gain, more efficient than competing methods and easier to manage.

Verbeeck, P., D. Deloddere, and M. De Prycker, "Introduction Strategies Towards B-ISDN for Business and Residential Subscribers Based on ATM," *IEEE Journal on Selected Areas in Communications*, Vol. 10, No. 9, Dec. 1992, pp. 1427–1433. A generalized, evolutionary approach for how ATM-based future services will be introduced to the office and home environments.

Vetter, R. J., "ATM Concepts, Architectures, and Protocols," *Communications of the ACM*, Vol. 38, No. 2, Feb. 1995, pp. 30–38, 109. Excellent short introduction to ATM and its extant problems as of 2/95. Especially clear explanations of classes and AALs, LAN emulation, and why the "asynchronous" in ATM.

Vickers, B. J., and T. Suda, "Connectionless Service for Public ATM Networks," *IEEE Communications Magazine*, Aug. 1994, pp. 34–42. Sees the interconnection of connectionless LANs and MANs over a connection-oriented B-ISDN presenting a dilemma with regard to efficient interoperability. Presents a scheme through which best-effort connectionless service could be provided in public ATM networks using connectionless servers and hop-by-hop flow control.

Viniotis, Y., and R. O. Onvural, *Asynchronous Transfer Mode Networks*, New York, NY: Plenum Press, 1993. Book is a series of invited papers presented at Tricomm'93, held April 26–27, 1993, in Raleigh, NC. The conference was dedicated to issues relative to the deployment of ATM networks. Contains 17 conference papers, including the sparkling "Electropolitical Correctness and High-Speed Networking, Or Why ATM Is Like A Nose," by D. Stevenson.

Watson, G. C., and S. Tohmé, "S++ — A New MAC Protocol for Gbps Local Area Networks," *IEEE Journal on Selected Areas in Communications*, Vol. 11, No. 4, May 1993, pp. 531–539. Concludes that lack of knowledge of the networking requirements of high-performance applications means that it is very difficult to select the "best" protocol. The S++

is designed for Gbps LANs due to its ease of implementation and excellent performance for asynchronous traffic. A modification of S, it removes the dependency between the physical location of the node and the performance of that node, as well as allowing even a single node to use every slot on the network.

Williams, K. A., T. Q. Dam, and D. H.-C. Du, "A Media-Access Protocol for Time- and Wavelength- Division Multiplexed Passive Star Networks," *IEEE Journal on Selected Areas in Communications*, Vol. 11, No. 4, May 1993, pp. 560–567. Presents a new media access protocol for time- and wavelength- division multiplexed optical passive star networks based on a new "bus-mesh" virtual topology.

Wright, D. J., "Strategic Impact of Broadband Telecommunications in Insurance, Publishing, and Health Care," *IEEE Journal on Selected Areas in Communications*, Vol. 10, No. 9, Dec. 1992, pp. 1369–1381. Unimaginative, business school prognostication of what areas BISDN might be employed in insurance, medical, and publishing industries. Little acquaintance with or insight into those industries and their economics.

Wright, D. J., et al., "BISDN Applications and Economics," *IEEE Journal on Selected Areas in Communications*, Vol. 10, No. 9, Dec. 1992, pp. 1365–1368. Reviews as of 12/92 the status of frame relay, SMDS/IEEE 802.6, SONET/SDH, and ATM and views current products as compliant, interfacable, or evolvable. Also, poses, with 622-Mbps links connecting distributed supercomputers, the prospects of the "backplane of a meta-computer."

Wright, D., Broadband, *Business Services, Technologies, and Strategic Impact*, Norwood, MA: Artech House, 1993. A business-oriented textbook whose message is that information is a strategic resource. Directed toward tomorrow's business leadership and business education (engineering management, MBA, MIS), it overviews broadband technologies (FR, SONET, DQDB, ATM), business services (SMDS, kinds of data services, service descriptions, multimedia) and potential strategic impacts on selected industries (insurance, publishing, healthcare) with some brief case examples.

Wu, T.-H., D. T. Kong, and R. C. Lau, "An Economic Feasibility Study for a Broadband Virtual Path SONET/ATM Self-Healing Ring Architecture," *IEEE Journal on Selected Areas in Communications*, Vol. 10, No. 9, Dec. 1992, pp. 1459–1473. Paper is an economic analysis of the feasibility of using ATM VP-based technology (called SONET ATM ring point-to-point

virtual path or SARPVP) to reduce the SONET ring cost of supporting cus-
tomer-demanded DS-1 service. The SARPVP architecture is essentially a
compromise of SONET/STM and ATM architecture: it takes the system
simplicity from SONET/STM and the flexibility from ATM.

Wu, T.-H., "Cost-Effective Network Evolution," *IEEE Communications
Magazine*, Sept. 1993, pp. 64–73. A proposal to migrate present synchronous transfer mode (STM) SONET rings to a SONET/ATM ring architecture using point-to-point virtual paths (SARPVP). Contends that SARPVP
will provide a much more efficient and cost-effective transport for bursty
high-speed data services than SONET/ATM due to the inherent charac-
teristic of ATM technology to share bandwidth and the simple add-drop
mux design for the SARPVP architecture.

Yang, S.-C., and J. A. Silvester, "A Reconfigurable ATM Switch Fabric for
Fault Tolerance and Traffic Balancing, *IEEE Journal on Selected Areas in
Communications*, Vol. 9, No. 8, Oct. 1991, pp. 1205–1217. Authors propose a large-scale ATM switch fabric where fault tolerance is achieved by
dynamic reconfiguration of the module interconnection network, and this
reconfiguration capability can also be used to ameliorate imbalanced traffic flows.

Yoshikai, N., and T.-H. Wu, "Control Protocol and Its Performance Analysis for Distributed ATM Virtual Path Self-Healing Network," *IEEE Journal
on Selected Areas in Communications*, Vol. 12, No. 6, Aug. 1994,
pp. 1020–1030. Simulates restoration in a metro LATA model network
with 15 nodes and 28 links with ATM VPs (AVPs). Believes AVPs to be
superior to SONET digital cross-connects for self-healing (dynamic) networks. Simulation implies that total LATA restoration time with AVPs
would be less than one second.

Zafirovic-Vukotic, M., and I. G. Niemegeers, "Multimedia Communications Systems: Upper Layers in the OSI Reference Model," *IEEE Journal
on Selected Areas in Communications*, Vol. 10, No. 9, Dec. 1992, pp.
1397–1402. Discusses upper level protocols and their suitability for multimedia processing. Sees that multimedia will require additional services,
particularly mutual stream synchronization, integral management, and
multiparty connections, that the subnets will not provide.

Zegura, W. W., "Architectures for ATM Switching Systems," *IEEE Communications Magazine*, Vol. 31, No. 2, Feb. 1993, pp. 28–37. The author
tries to predict the cost efficiency of ATM architectures based on estimated chip counts. Unfortunately, she excludes I/O, the integration of

multiple planes, and multicast ability in her estimates, and, given the custom design integration capabilities of today, it's hard to take her methodology and her results seriously, even as a contributing factor in making decisions on switch architecture. However, her characterization of the problem (eight candidate architectures) and her references and overview are useful.

Zitterbart, M., B. Stiller, and A. Tantawy, "A Model for Flexible High-Performance Communication Subsystems," *IEEE Journal on Selected Areas in Communications*, Vol. 11, No. 4, May 1993, pp. 507–518. A powerful indictment of the status quo in communications. Believes that (1) applications need efficient communications systems tailored to their individual needs; (2) that network bandwidth is increasing at least one order of magnitude faster than processor performance; (3) that strict hierarchical layering, typical of traditional communications models, limits the ability of mapping protocol entity tasks on parallel platforms since it implies sequential processing of layers; (4) layered communications models such as ISO limit efficiency in modern high-performance communications environments; (5) believes there is a need for automated creation of protocol machines to meet individual application requirements; and (6) the CCITT classification of telecommunications services is not very useful from communications subsystem design point of view.

Acronym List

(This collection is a superset of the terms collected by Donald R. Marks and Zelda Ford for Data Communications magazine and those posted by the ATM Forum in http://www.atmforum.com/.)

AAL   ATM adaptation layer: The standards layer that allows multiple applications to have data converted to and from the ATM cell. A protocol used that translates higher layer services into the size and format of an ATM cell.

AAL connection   Association established by the AAL between two or more next higher layer entities.

AAL-1   ATM adaptation layer type 1: AAL functions in support of constant bit rate, time-dependent traffic such as voice and video.

AAL-2   ATM adaptation layer type 2: This AAL is still undefined by the International Standards bodies. It is a placeholder for variable bit rate video transmission.

AAL-3/4   ATM adaptation layer type 3/4: AAL functions in support of variable bit rate, delay-tolerant data traffic requiring some sequencing and/or error detection support. Originally two AAL types, i.e. connection-oriented and connectionless, which have been combined.

AAL-5   ATM adaptation layer type 5: AAL functions in support of variable bit rate, delay-tolerant connection-oriented data traffic requiring minimal sequencing or error detection support.

ABR   Available bit rate: ABR is an ATM layer service category for which the limiting ATM layer transfer characteristics provided by the network may change subsequent to connection establishment. A flow control mechanism is specified that supports several types of feedback to control the source rate in response to changing ATM layer transfer characteristics. It is expected that an end system that adapts its traffic in accordance with the feedback will experience a low cell loss ratio and obtain a fair share of the available bandwidth according to a network-specific allocation policy. Cell delay variation is not controlled in this service, although admitted cells are not delayed unnecessarily.

ACM   Address complete message: A BISUP call control message from the receiving exchange to sending exchange indicating the completion of address information.

ACR   Allowed cell rate: An ABR service parameter, ACR is the current rate in cells/sec at which a source is allowed to send.

ACR   Attenuation to crosstalk ratio: One of the factors that limits the distance a signal may be sent through a given media. ACR is the ratio of the power of the received signal, attenuated by the media, over the power of the NEXT crosstalk from the local transmitter, usually expressed in decibels (dB). To achieve a desired bit error rate, the received signal power must usually be several times larger than the NEXT power or plus several decibels. Increasing a marginal ACR may decrease the bit error rate.

ACT   Activity bit

Address prefix   A string of 0 or more bits up to a maximum of 152 bits that is the lead portion of one or more ATM addresses.

Address resolution   Address resolution is the procedure by which a client associates a LAN destination with the ATM address of another client or the BUS.

Adjacency   The relationship between two communicating neighboring peer nodes.

Administrative domain   A collection of managed entities grouped for administrative reasons.

ADPCM   Adaptive differential pulse code modulation: A reduced bit rate variant of PCM audio encoding (see also PCM). This algorithm encodes the difference between an actual audio sample amplitude and a predicted amplitude and adapts the resolution based on recent differential values.

ADTF ACR decrease time factor: This is the time permitted between sending RM-cells before the rate is decreased to ICR (initial cell rate). The ADTF range is .01 to 10.23 sec with granularity of 10 ms.

AFI Authority and format identifier: This identifier is part of the network-level address header.

Aggregation token A number assigned to an outside link by the border nodes at the ends of the outside link. The same number is associated with all uplinks and induced uplinks associated with the outside link. In the parent and all higher level peer groups, all uplinks with the same aggregation token are aggregated.

AHFG ATM-attached host functional group: The group of functions performed by an ATM-attached host that is participating in the MPOA service.

Ai Signaling ID assigned by exchange A.

AIM ATM inverse multiplexer: A term discontinued because of conflict with an established product. Refer to AIMUX.

AIMUX ATM inverse multiplexing: A device that allows multiple T1 or E1 communications facilities to be combined into a single broadband facility for the transmission of ATM cells.

AIR Additive increase rate: An ABR service parameter, AIR controls the rate at which the cell transmission rate increases. It is signaled as AIRF, where AIRF = AIR * Nrm/PCR.

AIRF Additive increase rate factor: Refer to AIR.

AIS Alarm indication signal: An all 1's signal sent downstream or upstream by a device when it detects an error condition or receives an error condition or receives an error notification from another unit in the transmission path.

AIS-E Alarm indication signal external

AL 1 circuit emulation modes   ATM adaptation layer 1 (AAL 1), which handles constant bit rate (CBR) traffic, can emulate time division multiplexer (TDM) circuits in one of two ways. One of those ways is called the synchronous residual time stamp (SRTS) (or unstructured data transfer [UDT] mode). The other is called the structured data transfer (SDT) mode. (For more about SRTS and SDT, see below.)

AL service modes   There are two ways for ATM-aware protocols or native ATM APIs (application program interfaces) to pass data to an AAL for transmission across the ATM network: message mode and stream mode. In message mode, the AAL accepts an entire packet from a higher layer protocol or API and segments it into cell payloads before forwarding any cells onto the network. This can be an efficient way of sending information because it ensures only complete data packets are sent and reduces the need for retransmission. It also can introduce cell delay, particularly in application environments where large packets are used. In stream mode, the AAL begins transmitting cells as soon as payloads are filled, without waiting to receive an entire higher layer packet. This mode is faster but may be less efficient; it could result in more errored transmissions.

Alternate routing   A mechanism that supports the use of a new path after an attempt to set up a connection along a previously selected path fails.

AMI   Alternate mark inversion: A line coding format used on T1 facilities that transmits 1's by alternate positive and negative pulses.

Ancestor node   A logical group node that has a direct parent relationship to a given node (i.e., it is the parent of that node or the parent's parent,...).

ANI   Automatic number identification: A charge number parameter that is normally included in the initial address message to the succeeding carrier for billing purposes.

ANM   Answer message: A BISUP call control message from the receiving exchange to the sending exchange indicating answer and that a through-connection should be completed in both directions.

ANSI   American National Standards Institute: A U.S. standards body.

API   Application program interface: API is a programmatic interface used for interprogram communications or for interfacing between protocol layers.

API_connection Native ATM application program interface connection: API_connection is a relationship between an API_endpoint and other ATM devices that has the following characteristics: Data communication may occur between the API_endpoint and the other ATM devices comprising the API_connection. API_connection may occur over a duration of time only once; the same set of communicating ATM devices may form a new connection after a prior connection is released. The API_connection may be presently active (able to transfer data) or merely anticipated for the future

APPN Advanced peer-to-peer network: IBM network architecture for building dynamic routing across arbitrary network topologies. Intended as an eventual replacement for SNA, IBM's static routed, hierarchical network architecture.

ARE All routes explorer: A specific frame initiated by a source that is sent on all possible routes in source route bridging.

ARP Address resolution protocol: The procedures and messages in a communications protocol that determine which physical network address (MAC) corresponds to the IP address in the packet.

ARQ Automated repeat request

ASE Application service element

ASIC Application-specific integrated circuit

ASN Abstract syntax notation

ASP Abstract service primitive: An implementation-independent description of an interaction between a service user and a service provider at a particular service boundary, as defined by Open Systems Interconnection (OSI).

Assigned cell Cell that provides a service to an upper layer entity or ATM layer management entity (ATMM-entity).

ATD Asynchronous time division

Asynchronous time division multiplexing A multiplexing technique in which a transmission capability is organized in a priori unassigned time slots. The time slots are assigned to cells upon request of each application's instantaneous real need.

Asynchronous transfer mode (ATM)   A high-speed, connection-oriented switching and multiplexing technology that uses 53-byte cells (5-byte header, 48-byte payload) to transmit different types of traffic simultaneously, including voice, video, and data. It is asynchronous in that information streams can be sent independently without a common clock. ATM can be described logically in three planes: the user plane coordinates the interface between user protocols, such as IP or SMDS and ATM; the management plane coordinates the layers of the ATM stack; the control plane coordinates signaling and setting up and tearing down virtual circuits.

ATM adaptation layer (AAL)   A set of four standard protocols that translate user traffic from the higher layers of the protocol stack into a size and format that can be contained in the payload of an ATM cell and return it to its original form at the destination. Each AAL consists of two sublayers: the segmentation and reassembly (SAR) sublayer and the convergence sublayer. Each is geared to a particular class of traffic, with specific characteristics concerning delay and cell loss. All AAL functions occur at the ATM end station rather than at the switch. AAL 1 addresses CBR (constant bit rate) traffic such as digital voice and video and is used for applications that are sensitive to both cell loss and delay and to emulate conventional leased lines. It requires an additional byte of header information for sequence numbering, leaving 47 bytes for payload. AAL 2 is used with time-sensitive VBR (variable bit rate) traffic such as packetized voice. It allows ATM cells to be transmitted before the payload is full to accommodate an application's timing requirements. The AAL 2 spec has not been completed by the ATM Forum. AAL 3/4 handles bursty connection-oriented traffic, like error messages, or variable rate connectionless traffic, like LAN file transfers. It is intended for traffic that can tolerate delay but not cell loss; to ensure that the latter is kept to a minimum, AAL 3/4 performs error detection on each cell and uses a sophisticated error-checking mechanism that consumes 4 bytes of each 48-byte payload. AAL 3/4 allows ATM cells to be multiplexed. AAL 5 accommodates bursty LAN data traffic with less overhead than AAL 3/4. Also known as the simple and efficient adaptation layer (SEAL), AAL 5 uses a conventional 5-byte header. It does not support cell multiplexing.

ATM adaptation layer 0 (AAL 0)   Sometimes referred to as the undefined AAL, this is simply ATM with no convergence sublayer. Like other AALs, which establish an interface between the ATM layer and higher level protocols and interfaces, AAL 0 consists of a convergence sublayer and a segmentation and reassembly sublayer. Unlike other AALs, AAL 0 does not define a service-specific convergence sublayer (SSCS) or a protocol data unit (PDU), so it can only handle standard 53-byte ATM cells. In the future, AAL 0 may be used to transfer ATM cell data directly between the ATM network and high-level applications for APIs (application program interfaces).

ATM address Defined in the UNI specification as three formats, each having 20 bytes in length, including country, area, and end-system identifiers.

ATM address At 20 bytes long, ATM addresses scale to very large networks. Addressing is hierarchical, as in a phone network, using prefixes similar to area codes and exchanges. ATM switches share address information with attached end stations and maintain end-station addresses in routing tables. ATM source and destination addresses are not included within each cell, but are used by the switches to establish virtual path and virtual channel identifiers (VPIs/VCIs). This is in marked contrast to some LAN protocols, like TCP/IP and IPX, which pack destination addresses into each packet.

ATM API (application program interface) Although several vendors have written proprietary code, no standard ATM API yet exists. The ATM Forum is working on an API that will let application developers take advantage of ATM's quality of service and traffic management features; some Forum members are pushing for a single API that supports other network interfaces when ATM is not available.

ATM CSU/DSU (channel/data service unit) Segments ATM-compatible information, such as DXI (data exchange interface) frames generated by a router into ATM cells and then reassembles them at their destination.

ATM Forum The primary organization developing and defining ATM standards. Principal members participate in committees and vote on specifications; auditing members cannot participate in committees, but receive technical marketing documentation; user members participate only in end-user roundtables. Formed in 1991 by Adaptive Corp. (Redwood City, CA), Cisco Systems Inc. (San Jose, CA), Northern Telecom Ltd. (Mississauga, Ontario), and Sprint Corp. (Kansas City, MO), the Forum currently consists of 606 manufacturers, carriers, end users, and other interested parties.

ATM layer link   A section of an ATM layer connection between two adjacent active ATM layer entities (ATM-entities).

ATM layer   The layer of the ATM protocol stack that handles most of the processing and routing activities. These include building the ATM header, cell multiplexing/demultiplexing, cell reception and header validation, cell routing using VPIs/VCIs, payload-type identification, quality of service specification, and flow control and prioritization.

ATM link   A virtual path link (VPL) or a virtual channel link (VCL).

ATM management objects (IETF RFC 1695)   Proposed by the AToM MIB working group of the Internet Engineering Task Force (IETF), this specification defines objects used for managing ATM devices, networks, and services. ATM management objects allow net managers to consider groups of switches, virtual connections, interfaces, and services as discrete entities. The specification also specifies an MIB module for storing information about managed objects that complies with the semantics of SNMP versions 1 and 2.

ATMARP   ATM address resolution protocol

ATOM   ATM output buffer modular

ATM peer-to-peer connection   A virtual channel connection (VCC) or a virtual path connection (VPC).

ATM primitives   These are the various function calls and associated parameters that are possible on two adjacent layers of the ATM stack. For example, an ATM adaptation layer (AAL) requesting a particular class of service (constant bit rate, variable bit rate, etc.) must send a service request or service data unit (SDU) to the ATM layer of the network. This is accomplished by means of the ATM-DATA.request primitive. The ATM layer may also alert the AAL above it that a particular class of service is required. This is done by using the ATM-DATA.indication primitive.

ATM traffic descriptor   A generic list of traffic parameters that can be used to capture the intrinsic traffic characteristics of a requested ATM connection.

ATM user-user connection   An association established by the ATM layer to support communication between two or more ATM service users (i.e., between two or more next higher entities or between two or more ATM-entities). The communications over an ATM layer connection may be either bidirectional or unidirectional. The same virtual channel identifier (VCI) is issued for both directions of a connection at an interface.

ATM Asynchronous transfer mode: A transfer mode in which the information is organized into cells. It is asynchronous in the sense that the recurrence of cells containing information from an individual user is not necessarily periodic.

ATS Abstract test suite: A set of abstract test cases for testing a particular protocol. An "executable" test suite may be derived from an abstract test suite.

Attenuation The process of the reduction of the power of a signal as it passes through most media. Usually proportional to distance, attenuation is sometimes the factor that limits the distance a signal may be transmitted through a media before it can no longer be received.

Audiovisual multimedia services (AMS) Specifies service requirements and defines application program interfaces (APIs) for broadcast video, videoconferencing, and multimedia traffic. AMS is being developed by the ATM Forum's Service Aspects and Applications (SAA) working group. An important debate in the SAA concerns how MPEG-2 applications will travel over ATM. Early developers chose to carry MPEG-2 over ATM adaptation layer 1 (AAL 1); others found AAL 5 a more workable solution. Recently, some have suggested coming up with a new video-only AAL using the still-undefined AAL 2. At the moment, there is no consensus on which approach will become the standard.

AUU ATM user to user

Autoprovisioning Refers to automated setup, configuration, and management of permanent virtual circuits (PVCs). Supported by many carrier class switches, autoprovisioning enables net managers to establish a PVC simply by specifying the endpoints of the connection. In some cases, autoprovisioning also permits dynamic routing between circuits for load balancing and fault management. This is not the same as "soft" or "smart" PVCs, which allow users to set up connections to specific PVCs within a carrier or backbone using dial-up or ISDN circuits. Nor is it the same as switched virtual circuits (SVCs), which permit individual applications to set up and tear down connections as needed.

Available bit rate (ABR)   A class of service in which the ATM network makes a "best effort" to meet the traffic's bit rate requirements. ABR requires the transmitting end station to assume responsibility for data that cannot get through and does not guarantee delivery.

AVP   ATM virtual path

BBC   Broadband bearer capability: A bearer class field that is part of the initial address message.

B-ICI   Broadband intercarrier interface

B-ICI SAAL   B-ICI signaling ATM adaptation layer: A signaling layer that permits the transfer of connection control signaling and ensures reliable delivery of the protocol message. The SAAL is divided into a service-specific part and a common part (AAL5).

B-ICI   B-ISDN intercarrier interface: An ATM Forum–defined specification for the interface between public ATM networks to support user services across multiple public carriers.

B-ISDN   Broadband ISDN: A high-speed network standard (above 1.544 Mbps) that evolved narrowband ISDN with existing and new services with voice, data, and video in the same network.

B-ISUP   Broadband ISDN user's part

B-LLI   Broadband low layer information: This is a Q.2931 information element that identifies a layer 2 and a layer 3 protocol used by the application.

BCBDS   Broadband connectionless data bearer service

BCD   Binary coded decimal: A form of coding of each octet within a cell where each bit has one of two allowable states, 1 or 0.

BCOB   Broadband class of bearer

BCOB-A   Bearer class A: Indicated by ATM end user in SETUP message for connection-oriented, constant bit rate service. The network may perform internetworking based on AAL information element (IE).

BCOB-C Bearer class C: Indicated by ATM end user in SETUP message for connection-oriented, variable bit rate service. The network may perform internetworking based on AAL information element (IE).

BCOB-X Bearer class X: Indicated by ATM end user in SETUP message for ATM transport service where AAL, traffic type and timing requirements are transparent to the network.

BCOB Broadband connection oriented bearer: Information in the SETUP message that indicates the type of service requested by the calling user.

BECN Backward explicit congestion notification: A resource management (RM) cell type generated by the network or the destination, indicating congestion or approaching congestion for traffic flowing in the direction opposite that of the BECN cell.

BER Bit error rate: A measure of transmission quality. It is generally shown as a negative exponent, (e.g., 10-7 which means 1 out of 107 bits are in error or 1 out of 10,000,000 bits are in error).

BGP Border gateway protocol

BGT Broadcast and group translators

BHLI Broadband high layer information: This is a Q.2931 information element that identifies an application (or session layer protocol of an application).

Bi Signaling ID assigned by exchange B.

BIP Bit interleaved parity: A method used at the PHY layer to monitor the error performance of the link. A check bit or word is sent in the link overhead covering the previous block or frame. Bit errors in the payload will be detected and may be reported as maintenance information.

BIS Border intermediate system

BISDN Broadband integrated services digital network

BISSI Broadband interswitching system interface

BISUP   Broadband ISDN user's Part: An SS7 protocol that defines the signaling messages to control connections and services.

BN   BECN cell: A resource management (RM) cell type indicator. A backwards explicit congestion notification (BECN) RM-cell may be generated by the network or the destination. To do so, BN = 1 is set to indicate the cell is not source-generated, and DIR = 1 to indicate the backward flow. Source-generated RM-cells are initialized with BN = 0.

BN   Bridge number: A locally administered bridge ID used in source route bridging to uniquely identify a route between two LANs.

B-NT   Broadband network termination

BOF   Birds of a feather

BOM   Beginning of message: An indicator contained in the first cell of an ATM segmented packet.

BOOTP   Bootstrap protocol

Border node   A logical node that is in a specified peer group, and has at least one link that crosses the peer group boundary.

BOSS   Broadband overlay operations system

BPDU   Bridge protocol data unit: A message type used by bridges to exchange management and control information.

BPP   Bridge port pair (source routing descriptor): Frame header information identifying a bridge/LAN pair of a source route segment.

BPS   Bits per second

Broadband access   An ISDN access capable of supporting one or more broadband services.

Broadband integrated services digital network (B-ISDN)   A technology suite geared to multimedia. There are two transmission schemes: STM (synchronous transfer mode) and ATM.

Broadband intercarrier interface (BICI) A carrier-to-carrier interface similar to PNNI (private network-to-network interface) but lacking some of the detailed information offered by the latter. The difference arises because carriers are less likely to let their switches share routing information or detailed network maps with their competitors' gear. BICI now supports only PVCs (permanent virtual circuits) between carrier networks; the ATM Forum's BICI working group is currently addressing SVCs (switched virtual circuits).

Broadband A service or system requiring transmission channels capable of supporting rates greater than the integrated services digital network (ISDN) primary rate.

Broadcast Data transmission to all addresses or functions.

BSVC Broadcast switched virtual connections

BT Burst tolerance: BT applies to ATM connections supporting VBR services and is the limit parameter of the GCRA.

Btag Beginning tag: A 1-octet field of the CPCS_PDU used in conjunction with the Etag octet to form an association between the beginning of message and end of message.

B-TE Broadband terminal equipment: An equipment category for B-ISDN that includes terminal adaptors and terminals.

BUS Broadcast and unknown server: This server handles data sent by an LE client to the broadcast MAC address ('FFFFFFFFFFFF'), all multicast traffic, and initial unicast frames that are sent by a LAN emulation client.

BW Bandwidth: A numerical measurement of throughput of a system or network.

CA Cell arrival

CAC Call acceptance control

CAC Connection admission control: Connection admission control is defined as the set of actions taken by the network during the call setup phase (or during call renegotiation phase) in order to determine whether a connection request can be accepted or should be rejected (or whether a request for reallocation can be accommodated).

Call A call is an association between two or more users or between a user and a network entity that is established by the use of network capabilities. This association may have zero or more connections.

CAS Channel-associated signaling: A form of circuit state signaling in which the circuit state is indicated by one or more bits of signaling status sent repetitively and associated with that specific circuit.

CBDS Connectionless broadband data service: A connectionless service similar to Bellcore's SMDS defined by European Telecommunications Standards Institute (ETSI).

CBR Constant bit rate: An ATM service category that supports a constant or guaranteed rate to transport services such as video or voice as well as circuit emulation, which requires rigorous timing control and performance parameters.

CBR interactive Constant bit rate interactive

CBR noninteractive Constant bit rate noninteractive

CC Continuity cell

CCITT Consultative Committee on International Telephone & Telegraph

CCR Current cell rate: The current cell rate is an RM-cell field set by the source to its current ACR when it generates a forward RM-cell. This field may be used to facilitate the calculation of ER and may not be changed by network elements. CCR is formatted as a rate.

CCS Common channel signaling: A form signaling in which a group of circuits share a signaling channel. Refer to SS7.

CD-ROM Compact disk-read only memory: Used by a computer to store large amounts of data. Commonly used for interactive video games.

CDF Cutoff decrease factor: CDF controls the decrease in ACR (allowed cell rate) associated with CRM.

CDT   Cell delay tolerance

CDV   Cell delay variation: CDV is a component of cell transfer delay, induced by buffering and cell scheduling. Peak-to-peak CDV is a QoS delay parameter associated with CBR and VBR services. The peak-to-peak CDV is the ((1 - a) quantile of the CTD) minus the fixed CTD that could be experienced by any delivered cell on a connection during the entire connection holding time. The parameter a is the probability of a cell arriving late. See CDVT.

CDVT   Cell delay variation tolerance: ATM layer functions may alter the traffic characteristics of ATM connections by introducing cell delay variation. When cells from two or more ATM connections are multiplexed, cells of a given ATM connection may be delayed while cells of another ATM connection are being inserted at the output of the multiplexer. Similarly, some cells may be delayed while physical layer overhead or OAM cells are inserted. Consequently, some randomness may affect the inter-arrival time between consecutive cells of a connection as monitored at the UNI. The upper bound on the "clumping" measure is the CDVT.

CE   Connection endpoint: A terminator at one end of a layer connection within a SAP.

CEI   Connection endpoint identifier: Identifier of a CE that can be used to identify the connection at a SAP.

Cell delay variation (CDV)   Measures the allowable variance in delay between one cell and the next, expressed in fractions of a second. When emulating a circuit, CDV measurements allow the network to determine if cells are arriving too fast or too slow.

Cell header   ATM Layer protocol control information.

Cell interarrival variation (CIV)   "Jitter" in common parlance, CIV measures how consistently ATM cells arrive at the receiving end station. Cell interarrival time is specified by the source application and should vary as little as possible. For constant bit rate (CBR) traffic, the interval between cells should be the same at the destination and the source. If it remains constant, the latency of the ATM switch or the network itself (also known as cell delay) will not affect the cell interarrival interval. But if latency varies, so will the interarrival interval. Any variation could affect the quality of voice or video applications.

Cell multiplexing/demultiplexing   An ATM layer function that groups cells belonging to different virtual paths or circuits and transmits them in a stream to the target switch, where they are demuxed and routed to the correct end-points.

Cell rate decoupling   Technique used by the ATM network to make all data transmissions correspond to the interface cell rate. To achieve an application's required data rate and match the speed of the ATM interface, the network inserts idle cells (cells with no payload content) between valid cells. For example, if a videoconferencing application requires an 8-Mbps virtual circuit (VC) and is the only application transmitting across a 155-Mbps interface, the ATM network will transmit enough idle cells to fill the 155-Mbps pipe while delivering application data in a separate VC at 8 Mbps. Idle cells are discarded at the receiving station.

Cell transfer outcomes   A set of performance characteristics used to track what happens to cells transmitted across an ATM network and to evaluate network performance. Possible outcomes include (1) successful cell transfer: The cell is received within the required timeframe and it contains correct payload information and a valid header field; (2) errored cell transfer: the cell is received within the specified time, but the binary content of the payload differs from what the application requires or the cell has an invalid header field; (3) lost cell outcome: no cell is received within the specified time (this applies to cells that arrive too late and to those that never arrive); (4) misinserted cell transfer: the cell arrives on time but does not correspond to a cell that was sent by the application (typically, a misinserted cell appears because of a cell header error); (5) severely errored cell block transfer: A block of errored, misinserted, or lost cells. The number or percentage of cells making up a severely errored block depends on the user application.

Cell loss priority (CLP) field   A priority bit in the cell header; when set, it indicates that the cell can be discarded if necessary.

Cell   A unit of transmission in ATM. A fixed-size frame consisting of a 5-octet header and a 48-octet payload.

Cell   An ATM cell consists of 53 bytes or "octets." Of these, 5 constitute the header; the remaining 48 carry the data payload.

CER   Cell error ratio: The ratio of errored cells in a transmission in relation to the total cells sent in a transmission. The measurement is taken over a time interval and it is desirable to be measured on an in-service circuit.

CES Circuit emulation service: The ATM Forum circuit emulation service interoperability specification specifies interoperability agreements for supporting constant bit rate (CBR) traffic over ATM networks that comply with the other ATM Forum interoperability agreements. Specifically, this specification supports emulation of existing TDM circuits over ATM networks.

Child node A node at the next lower level of the hierarchy that is contained in the peer group represented by the logical group node currently referenced. This could be a logical group node or a physical node.

Child peer group A child peer group of a peer group is any containing a child node of a logical group node in that peer group. A child peer group of a logical group node is the one containing the child node of that logical group node.

CI Congestion indicator: This is a field in an RM-cell and is used to cause the source to decrease its ACR. The source sets CI = 0 when it sends an RM-cell. Setting CI = 1 is typically how destinations indicate that EFCI has been received on a previous data cell.

CIF Cells in flight: An ABR service parameter, CIF is the negotiated number of cells that the network would like to limit the source to sending during idle startup period, before the first RM-cell returns. Range: 0–16,777,215

CIF Cells in frames: CIF defines a way to encapsulate ATM cells in Ethernet and token ring frames, allowing ATM data to be transported to legacy desktops. The advantage to the scheme is that it enables developers to create ATM applications that run over hybrid legacy networks.

CIP Carrier identification parameter: A 3- or 4-digit code in the initial address message identifying the carrier to be used for the connection.

CIR Committed information rate: CIR is the information transfer rate that a network offering frame relay services (FRS) is committed to transfer under normal conditions. The rate is averaged over a minimum increment of time.

CL   Connectionless service: A service that allows the transfer of information among service subscribers without the need for end-to-end establishment procedures.

Classical IP and ARP over ATM   An adaptation of TCP/IP and its address resolution protocol (ARP) for ATM defined by the IETF (Internet Engineering Task Force) in RFCs (Requests for Comment) 1483 and 1577. It places packets and ARP requests directly into PDUs (protocol data units) and converts them into ATM cells. Classical IP does not recognize conventional MAC-layer protocols like Ethernet or token ring.

CLNAP   Connectionless network access protocol

CLNP   Connectionless network protocol

CLNS   Connectionless network service

CLP   Cell loss priority: This bit in the ATM cell header indicates two levels of priority for ATM cells. CLP = 0 cells are higher priority than CLP = 1 cells. CLP = 1 cells may be discarded during periods of congestion to preserve the CLR of CLP = 0 cells.

CLR   Cell loss ratio: CLR is a negotiated QoS parameter and acceptable values are network-specific. The objective is to minimize CLR provided the end system adapts the traffic to the changing ATM layer transfer characteristics. The cell loss ratio is defined for a connection as: lost cells/total transmitted cells. The CLR parameter is the value of CLR that the network agrees to offer as an objective over the lifetime of the connection. It is expressed as an order of magnitude, having a range of 10–1 to 10–15, and unspecified.

CLS   Connectionless server

CLSF   Connectionless service function

CME   Component management entity

CMI   Coded mark inversion

CMIP   Common management interface protocol: An ITU-TSS standard for the message formats and procedures used to exchange management information in order to operate, administer, maintain, and provision a network.

CMR   Cell misinsertion rate: The ratio of cells received at an endpoint that were not originally transmitted by the source end in relation to the total number of cells properly transmitted.

CN   Copy network

CNM   Customer network management

CNR   Complex node representation: A collection of nodal state parameters that provide detailed state information associated with a logical node.

CO   Connection-oriented

COD   Connection-oriented data: Data requiring sequential delivery of its component PDUs to assure correct functioning of its supported application, (e.g., voice or video).

COM   Continuation of message: An indicator used by the ATM adaptation layer to indicate that a particular ATM cell is a continuation of a higher layer information packet that has been segmented.

Common part convergence sublayer (CPCS)   The portion of the convergence sublayer of an AAL that remains the same regardless of the type of traffic.

Common peer group   The lowest level peer group in which a set of nodes is represented. A node is represented in a peer group either directly or through one of its ancestors.

Communication endpoint   An object associated with a set of attributes that are specified at the communication creation time.

Configuration   The phase in which the LE client discovers the LE service.

Congestion control   Mechanisms that control traffic flow so switches and end stations are not overwhelmed and cells dropped. ATM defines several simple schemes, among them GFC (generic flow control) and CLP fields in cell headers and the EFCI (explicit forward congestion indicator) bit in the PTI (payload type identifier). More sophisticated mechanisms are needed to deal with congestion in large ATM networks carrying different types of traffic. The ATM Forum recently ratified a rate-based traffic management strategy that counts on switches and end stations throttling back when congestion is encountered; a credit-based scheme also was considered, which relied more heavily on switch buffers. Other means of congestion control include UPC (usage parameter control) and CAC (connection admission control). When confronted by congestion, many ATM switches discard cells according to CLP. Since voice and video are not tolerant of cell loss, this can make it difficult to achieve quality of service parameters. Data traffic is more tolerant of loss and delay, but if cells containing information from a higher level packet are dropped, the entire packet may have to be transmitted. Considering that IP packets are 1,500 bytes long and FDDI packets 4,500 bytes, the loss of a single cell could cause significant retransmissions, further aggravating congestion.

Connection admission control (CAC) Two mechanisms used to control the set up of virtual circuits. Overbooking, which allows one connection to exceed permissible traffic limits, assumes that other active connections are not using the maximum available resources. Full booking limits network access once maximum resources are committed and only adds connections that specify acceptable traffic parameters.

Connection-oriented communications A form of cell switching or packet multiplexing characterized by individual virtual circuits based on virtual circuit identifiers. ATM is a connection-oriented technology.

Connection An ATM connection consists of concatenation of ATM layer links in order to provide an end-to-end information transfer capability to access points.

Connection In switched virtual connection (SVC) environments, the LAN emulation management entities set up connections between each other using UNI signaling.

connection. Each ABR control segment, except the last, is terminated by a virtual destination. A virtual destination assumes the behavior of an ABR destination endpoint. Forward RM-cells received by a virtual destination are turned around and not forwarded to the next segment of the connection.

Connection. Note that UNI 4.0 does not support this connection type.

Connectionless communications A form of cell switching or packet multiplexing that identifies individual channels based on global addresses rather than predefined virtual circuits. Used by shared-media LANs like FDDI and token ring.

Connectionless Refers to ability of existing LANs to send data without previously establishing connections.

Constant bit rate (CBR) Digital information, such as video and digitized voice, that must be represented by a continuous stream of bits. CBR traffic requires guaranteed throughput rates and service levels.

Control connections A control VCC links the LEC to the LECS. Control VCCs also link the LEC to the LES and carry LE_ARP traffic and control frames. The control VCCs never carry data frames.

Convergence sublayer (CS) The portion of the AAL that prepares information in a common format convergence sublayer-protocol data unit (CS-PDU) before it is segmented into cells and returns it to its original form after reassembly at the destination switch. At the destination switch, an ATM cell goes through reassembly in the SAR layer, then is passed up to the CS in the form of a CS-PDU. The CS then converts the PDU to whatever the traffic was in its original form.

Convergence sublayer-protocol data unit (CS-PDU) Information contained within a PDU that conforms to the specifications of the ATM convergence sublayer and is ready to be segmented into cells.

Corresponding entities Peer entities with a lower layer connection among them.

CoS Class of service

CP Connection processor

CPCS Common part convergence sublayer: The portion of the convergence sublayer of an AAL that remains the same regardless of the traffic type.

CPCS-SDU Common part convergence sublayer-service data unit: Protocol data unit to be delivered to the receiving AAL layer by the destination CP convergence sublayer.

CPE Customer premises equipment: End-user equipment that resides on the customer's premises, which may not be owned by the local exchange carrier.

CPG Call progress message

CPI Common part indicator

CPN Calling party number: A parameter of the initial address message that identifies the calling number and is sent to the destination carrier.

CPN customer premises network

Crankback A mechanism for partially releasing a connection setup in progress that has encountered a failure. This mechanism allows PNNI to perform alternate routing.

Crankback IE Crankback information element

CRC Cyclic redundancy check: A mathematical algorithm that computes a numerical value based on the bits in a block of data. This number is transmitted with the data and the receiver uses this information and the same algorithm to ensure the accurate delivery of data by comparing the results of algorithm and the number received. If a mismatch occurs, an error in transmission is presumed.

CRCG Common routing connection group

CRF Cell relay function: This is the basic function that an ATM network performs in order to provide a cell relay service to ATM end stations.

CRF Connection related function: A term used by traffic management to reference a point in a network or a network element where per-connection functions are occurring. This is the point where policing at the VCC or VPC level may occur.

CRF(VC) Virtual channel connection related function

CRF(VP) Virtual path connection related function

CRM Cell rate margin: This is a measure of the difference between the effective bandwidth allocation and the allocation for sustainable rate in cells per second.

CRM Missing RM-cell count: CRM limits the number of forward RM-cells that may be sent in the absence of received backward RM-cells.

CRS Cell relay service: A carrier service that supports the receipt and transmission of ATM cells between end users in compliance with ATM standards and implementation specifications.

CS Carrier selection

CS Convergence sublayer: The general procedures and functions that convert between ATM and non-ATM formats. This describes the functions of the upper half of the AAL layer. This is also used to describe the conversion functions between non-ATM protocols such as frame relay or SMDS and ATM protocols above the AAL layer.

CS1 Capability set one

CS2 Capability set two

CSI Convergence sublayer indication

CSPDN Circuit switched public data network

CS-PDU convergence sublayer-protocol data unit

CSR Cell missequenced ratio

CSU Channel service unit: An interface for digital leased lines that performs loopback testing and line conditioning.

CT Conformance test: Testing to determine whether an implementation complies with the specifications of a standard and exhibits the behaviors mandated by that standard.

CTD Cell transfer delay: This is defined as the elapsed time between a cell exit event at measurement point 1 (e.g., at the source UNI) and the corresponding cell entry event at measurement point 2 (e.g., the destination UNI) for a particular connection. The cell transfer delay between two measurement points is the sum of the total inter-ATM node transmission delay and the total ATM node processing delay.

CTV   Cell tolerance variation

Customer network management (CNM)   Allows users of ATM public networks to monitor and manage their portion of the carrier's circuits. Thus far, the ATM Forum has agreed that the CNM interface will give users the ability to monitor physical ports, virtual circuits, virtual paths, usage parameters, and quality of service parameters.

Cyclic redundancy check (CRC)   A mathematical algorithm used to en- sure accurate delivery based on the actual contents of the data.

DA   Destination address: Information sent in the forward direction indicating the address of the called station or customer.

DA   Destination MAC address: A 6-octet value uniquely identifying an endpoint and that is sent in IEEE LAN frame headers to indicate frame destination.

DAN   Desk area network

DAS   Dynamic allocation scheme

Data connections   Data VCCs connect the LECs to each other and to the broadcast and unknown server. These carry Ethernet/IEEE 802.3 or IEEE 802.5 data frames as well as flush messages.

Data exchange interface (DXI)   Defines a format for passing data that has gone through the ATM convergence sublayer (a CS-PDU) between a router and a CSU/DSU or other device with ATM SAR capability.

DBS   Direct broadcast services

DCC   Data country code: This specifies the country in which an address is registered. The codes are given in ISO 3166. The length of this field is 2 octets. The digits of the data country code are encoded in binary coded decimal (BCD) syntax. The codes will be left-justified and padded on the right with the hexadecimal value "F" to fill the two.

DCE Data communication equipment: A generic definition of computing equipment that attaches to a network via a DTE.

DD Depacketization delay

DDBS Developing distributed database system

D-D VCC Data-direct virtual circuit connection: A type of virtual circuit used to transfer between LAN emulation clients (LECs) in a LAN emulation (LANE) network. LANE also defines call-control VCCs, which are used to transfer call-control information between a LEC and a LES (LAN emulation server) or BUS (broadcast and unknown server).

Default node representation A single value for each nodal state parameter giving the presumed value between any entry or exit to the logical node and the nucleus. Defined with two parameters: the increment (I) and the limit (L).

Demultiplexing A function performed by a layer entity that identifies and separates SDUs from a single connection to more than one connection.

DES Destination end station: An ATM termination point that is the destination for ATM messages of a connection and is used as a reference point for ABR services. See SES.

DFI Domain-specific identifier

Dijkstra's algorithm An algorithm that is sometimes used to calculate routes given a link and nodal state topology database.

DIR This is a field in an RM-cell that indicates the direction of the RM-cell with respect to the data flow with which it is associated. The source sets DIR = 0 and the destination sets DIR = 1.

Direct set A set of host interfaces that can establish direct layer 2 communications for unicast (not needed in MPOA).

DLC Data link control

DLCI Data link connection identifier

DLPI UNIX International, Data Link Provider Interface (DLPI) Specification: Revision 2.0.0, OSI Work Group, August, 1991.

DMDD Distributed multiplexing distributed demultiplexing

DN Distribution network

Domain Refer to administrative domain.

DQDB Distributed queue dual bus

DRAM Dynamic random access memory

DS-0 Digital signal, level 0: The 64-Kbps rate that is the basic building block for both the North American and European digital hierarchies.

DS-1 Digital signal, level 1: The North American Digital Hierarchy signaling standard for transmission at 1.544 Mbps. This standard supports 24 simultaneous DS-0 signals. The term is often used interchangeably with T1 carrier although DS-1 signals may be exchanged over other transmission systems.

DS-2 Digital signal, level 2: The North American Digital Hierarchy signaling standard for transmission of 6.312 Mbps that is used by T2 carriers, which support 96 calls.

DS-3 Digital signal, level 3: The North American Digital Hierarchy signaling standard for transmission at 44.736 Mbps that is used by T3 carrier. DS-3 supports 28 DS-1s plus overhead.

DS Distributed single layer test method: An abstract test method in which the upper tester is located within the system under test and the point of control and observation (PCO) is located at the upper service boundary of the implementation under test (IUT)—for testing one protocol layer. Test events are specified in terms of the abstract service primitives (ASP) at the upper tester above the IUT and ASPs and/or protocol data units (PDU) at the lower tester PCO.

DS3 PLCP Physical layer convergence protocol: An alternate method used by older T carrier equipment to locate ATM cell boundaries. This method has recently been moved to an informative appendix of the ATM DS3 specification and has been replaced by the HEC method.

DSE Distributed single-layer embedded (test method): An abstract test method in which the upper tester is located within the system under test and there is a point of control and observation at the upper service boundary of the implementation under test (IUT) for testing a protocol layer, or sublayer, which is part of a multiprotocol IUT.

DSID Destination signaling identifier

DSS2 Setup digital subscriber signaling #2

DSS2 Setup Digital subscriber signaling #2: ATM broadband signaling.

DSU Data service unit: Equipment used to attach users' computing equipment to a public network.

DTE Data terminal equipment: A generic definition of external networking interface equipment such as a modem.

DTL Designated transit list: A list of nodes and optional link IDs that completely specify a path across a single PNNI peer group.

DTL Terminator: The last switching system within the entire PNNI routing domain to process the connection and thus the connection's DTL.

DTL IE DTL information element

DTL originator The first switching system within the entire PNNI routing domain to build the initial DTL stack for a given connection.

DXI Data exchange interface: A variable length frame-based ATM interface between a DTE and a special ATM CSU/DSU. The ATM CSU/DSU converts between the variable-length DXI frames and the fixed-length ATM cells.

DXI Data interchange interface

E.164 A public network addressing standard utilizing up to a maximum of 15 digits. ATM uses E.164 addressing for public network addressing.

E1 Also known as CEPT1, the 2.048-Mbps rate used by European CEPT carrier to transmit 30 64-Kbps digital channels for voice or data calls, plus a 64-Kbps signaling channel and a 64-Kbps channel for framing and maintenance.

E1   European standard for digital transmission service at 2.048 Mbps.

E3   Also known as CEPT3, the 34.368-Mbps rate used by European CEPT carrier to transmit 16 CEPT1s plus overhead.

E3   European standard for digital transmission service at 34.368 Mbps (transports 16 E1 circuits).

Early packet discard (EPD)   A congestion control technique that selectively drops all but the last ATM cell in a classical IP over ATM packet. When congestion occurs, EPD discards cells at the beginning of an IP packet, leaving the rest intact. The last cell is preserved because it alerts the switch and the destination station of the beginning of a new packet. Because IP packets from which cells have been discarded receive no acknowledgment from the destination, they are automatically retransmitted from the source. Most vendors expect EPD to be used in conjunction with UBR service. Switches simply junk UBR cells when congestion occurs, without regard for application traffic. By discarding cells selectively, so that whole IP packets are present, EPD makes UBR a safer option.

Edge device   A physical device that is capable of forwarding packets between legacy interworking interfaces (e.g., Ethernet, token ring, etc.) and ATM interfaces based on data link and network layer information but which does not participate in the running of any network layer routing protocol. An edge device obtains forwarding descriptions using the route distribution protocol.

EFCI   Explicit forward congestion indication: EFCI is an indication in the ATM cell header. A network element in an impending congested state or a congested state may set EFCI so that this indication may be examined by the destination end system. For example, the end system may use this indication to implement a protocol that adaptively lowers the cell rate of the connection during congestion or impending congestion. A network element that is not in a congestion state or an impending congestion state will not modify the value of this indication. Impending congestion is the state when a network equipment is operating around its engineered capacity level.

EFS Error free seconds: A unit used to specify the error performance of T carrier systems, usually expressed as EFS per hour, day, or week. This method gives a better indication of the distribution of bit errors than a simple bit error rate (BER). Also refer to SES.

EISA Extended industry standard architecture

ELAN Emulated local area network: A logical network initiated by using the mechanisms defined by LAN Emulation. This could include ATM and legacy attached end stations.

EMI Electromagnetic interference: Equipment used in high-speed data systems, including ATM, that generate and transmit many signals in the radiofrequency portion of the electromagnetic spectrum. Interference to other equipment or radio services may result if sufficient power from these signals escape the equipment enclosures or transmission media. National and international regulatory agencies (FCC, CISPR, etc.) set limits for these emissions. Class A is for industrial use and class B is for residential use.

EML Element management layer: An abstraction of the functions provided by systems that manage each network element on an individual basis.

EMS Element management system: A management system that provides functions at the element management layer.

End station These devices (e.g., hosts or PCs) enable the communication between ATM end stations and end stations on "legacy" LAN or among ATM end stations.

Entry border node The node that receives a call over an outside link. This is the first node within a peer group to see this call.

EOM End of message: An indicator used in the AAL that identifies the last ATM cell containing information from a data packet that has been segmented.

EPD Early packet discard

ER Explicit rate: The explicit rate is an RM-cell field used to limit the source ACR to a specific value. It is initially set by the source to a requested rate (such as PCR). It may be subsequently reduced by any network element in the path to a value that the element can sustain. ER is formatted as a rate.

ES End system: A system where an ATM connection is terminated or initiated. An originating end system initiates the ATM connection, and the terminating end system terminates the ATM connection. OAM cells may be generated and received.

ESF Extended superframe: A DS1 framing format in which 24 DS0 time slots plus a coded framing bit are organized into a frame that is repeated 24 times to form a superframe.

ESI End system identifier: This identifier distinguishes multiple nodes at the same level in case the lower level peer group is partitioned.

Etag End tag

ETE End to end

ETSI European Telecommunications Standards Institute: The primary telecommunications standards organization.

Exception A connectivity advertisement in a PNNI complex node representation that represents something other than the default node representation.

Exit border node The node that will progress a call over an outside link. This is the last node within a peer group to see this call.

EXM Exit message

Exterior link A link that crosses the boundary of the PNNI routing domain. The PNNI protocol does not run over an exterior link.

Exterior reachable address An address that can be reached through a PNNI routing domain, but which is not located in that PNNI routing domain.

Exterior route A route that traverses an exterior link.

Exterior Denotes that an item (e.g., link, node, or reachable address) is outside of a PNNI routing domain.

Fairness   As related to generic flow control (GFC), fairness is defined as meeting all the agreed quality of service (QoS) requirements by controlling the order of service for all active connections.

FC   Feedback control: Feedback controls are defined as the set of actions taken by the network and by the end systems to regulate the traffic submitted on ATM connections according to the state of network elements.

FCS   Fast circuit switching

FCS   Frame check sequence: Any mathematical formula that derives a numeric value based on the bit pattern of a transmitted block of information and uses that value at the receiving end to determine the existence of any transmission errors.

FDDI   Fiber Distributed Data Interface: A 100-Mbps local area network standard that was developed by ANSI that is designed to work on fiber-optic cables, using techniques similar to token ring.

FEBE   Far end block error: A maintenance signal transmitted in the PHY overhead that a bit error(s) has been detected at the PHY layer at the far end of the link. This is used to monitor bit error performance of the link.

FEC   Forward error correction: A technique for detection and correction of errors in a digital data stream.

FECN   Forward error control notification

FERF   Far end receive failure

FG   Functional group: A collection of functions related in such a way that they will be provided by a single logical component. Examples include the route server functional group (RSFG), the IASG (Internetwork address subgroup), the coordination functional group (ICFG), the edge device functional group (EDFG) and the ATM attached host behavior functional group (AHFG).

Flow control   See congestion control.

Flush protocol   The flush protocol is provided to ensure the correct order of delivery of unicast data frames.

Foreign address An address that does not match any of a given node's summary addresses.

Forwarding description The resolved mapping of an MPOA target to a set of parameters used to set up an ATM connection on which to forward packets.

FR Frame relay

FRS Frame relay service: A connection-oriented service that is capable of carrying up to 4,096 bytes per frame.

FR-SSCS Frame relay service-specific convergence sublayer

FRTT Fixed round-trip time: This is the sum of the fixed and propagation delays from the source to the furthest destination and back.

FUNI Frame user network interface

G.703 ITU-T Recommendation G.703, "Physical/Electrical Characteristics of Hierarchical Digital Interfaces."

G.704 ITU-T Recommendation G.704, "Synchronous Frame Structures Used at Primary and Secondary Hierarchy Levels."

G.804 ITU-T Recommendation G.804, "ATM Cell Mapping into Plesiochronous Digital Hierarchy (PDH)."

GAP Generic address parameter

GCAC Generic connection admission control: This is a process to determine if a link has potentially enough resources to support a connection.

GCID Global call identifier

GCID-IE Global call identifier-information element

GCRA Generic cell rate algorithm: The GCRA is used to define conformance with respect to the traffic contract of the connection. For each cell arrival, the GCRA determines whether the cell conforms to the traffic contract. The UPC function may implement the GCRA or one or more equivalent algorithms to enforce conformance. The GCRA is defined with two parameters: the increment (I) and the limit (L).

GDC General DataComm

Generic flow control (GFC) field Four priority bits in an ATM header. The default setting—four 0s—indicates that the cell is uncontrolled; thus it does not take precedence over another cell when contending for a virtual circuit. Setting any of the bits in the GFC field tells the target end station that the switch can implement some form of congestion control; the end station echoes this bit back to the switch to confirm that it can set priorities. The switch and end station could use the GFC field to prioritize voice over video, for example, or indicate that both voice and video take precedence over other types of data.

GFC Generic flow control: GFC is a field in the ATM header that can be used to provide local functions (e.g., flow control). It has local significance only and the value encoded in the field is not carried end to end.

GRAMS Gopher-style real-time ATM multimedia services

GRC Generic reference configuration

H-C H-channels are ISDN bearer services that have predefined speeds, starting and stopping locations on a PRI, and are contiguously transported from one PRI site through networks to another PRI site.

H0 channel A 384-Kbps channel that consists of six contiguous DS0s (64 Kbps) of a T1 line.

H10 channel The North American 1,472-Kbps channel from a T1 or primary rate carrier. This is equivalent to 23 64-Kbps channels.

H11 channel The North American primary rate used as a single 1,536-Kbps channel. This channel uses 24 contiguous DS0s or the entire T1 line except for the 8-Kbps framing pattern.

H12 The European primary rate used as a single 1,920-Kbps channel (30 64-Kbps channels or the entire E1 line except for the 64-Kbps framing and maintenance channel.

HBFG Host behavior functional group: The group of functions performed by an ATM-attached host that is participating in the MPOA service.

HDB3 High density bipolar 3

HDLC High level data link control: An ITU-TSS link layer protocol standard for point-to-point and multipoint communications.

Header error control (HEC) field A single byte containing the information needed for the transmission convergence (TC) sublayer of the ATM physical (PHY) layer to perform error detection on the cell header. If errors are found, the cell is dropped before processing moves up to the ATM layer where routing takes place.

Header Protocol control information located at the beginning of a protocol data unit.

Header The five bytes in an ATM cell that supply addressing and control information, including generic flow control, virtual path identifier, virtual circuit identifier, payload type, and cell loss priority.

HEC Header error check

HEC Header error control: Using the fifth octet in the ATM cell header, ATM equipment may check for an error and correct the contents of the header. The check character is calculated using a CRC algorithm allowing a single bit error in the header to be corrected or multiple errors to be detected.

HEL Header extension length

Hello packet A type of PNNI Routing packet that is exchanged between neighboring logical nodes.

Hierarchically complete source route A stack of DTLs representing a route across a PNNI routing domain such that a DTL is included for each hierarchical level between and including the current level and the lowest visible level in which the source and destination are reachable.

HiPPI High-performance protocol interface

HLPI   Higher layer protocol identifier

HOL   Head of line

Hop-by-hop route   A route that is created by having each switch along the path use its own routing knowledge to determine the next hop of the route, with the expectation that all switches will choose consistent hops such that the call will reach the desired destination. PNNI does not use hop-by-hop routing.

Horizontal link   A link between two logical nodes that belong to the same peer group.

Host apparent address   A set of internetwork layer addresses that a host will directly resolve to lower layer addresses.

HTDM   Hybrid TDM

I.356   ITU-T Specifications for traffic measurement.

I.361   B-ISDN ATM layer specification.

I.362   B-ISDN ATM layer (AAL) functional description.

I.363   B-ISDN ATM layer (AAL) specification.

I.432   ITU-T Recommendation for B-ISDN user-network interface.

IAA   Initial address acknowledgment

IAM   Initial address message

IAR   Initial address reject

IASG   Internetwork address subgroup: A range of internetwork layer addresses summarized in an internetwork layer routing protocol.

IC   Initial cell rate

ICD   International code designator: This identifies an international organization. The registration authority for the international code designator is maintained by the British Standards Institute. The length of this field is two octets.

ICMP   Internet control message protocol

ICR   Initial cell rate: An ABR service parameter, in cells/sec, that is the rate at which a source should send initially and after an idle period.

IDU   Interface data unit: The unit of information transferred to/from the upper layer in a single interaction across the SAP. Each IDU contains interface control information and may also contain the whole or part of the SDU.

IE   Information element

IEC   Interexchange carrier: A long distance telephone company.

IEEE 802.3   A local area network protocol suite commonly known as Ethernet. Ethernet has either a 10-Mbps or 100-Mbps throughput and uses carrier sense multiple access bus with collision detection CSMA/CD. This method allows users to share the network cable. However, only one station can use the cable at a time. A variety of physical medium dependent protocols are supported.

IEEE 802.5   A local area network protocol suite commonly known as token ring. A standard originated by IBM for a token-passing ring network that can be configured in a star topology. Versions supported are 4 Mbps and 16 Mbps.

IEEE   Institute of Electrical and Electronics Engineers: A worldwide engineering publishing and standards-making body for the electronics industry.

IETF   Internet Engineering Task Force: The organization that provides the coordination of standards and specification development for TCP/IP networking.

IISP   Interim interswitch signaling protocol

ILMI   Interim link management interface: An ATM Forum–defined interim specification for network management functions between an end user and a public or private network and between a public network and a private network. This is based on a limited subset of SNMP capabilities.

IMA Inverse multiplexing for ATM: This specification will enable ATM traffic to be inversely multiplexed across multiple carrier access lines. In essence, IMA defines a cyclic round-robin scheme for putting on circuits: the first cell is sent on the first circuit, the second on the second, and so on. With IMA, service providers can offer ATM services at rates between T1/E1 (1.54/2.048 Mbps), which are typically too slow for applications that need ATM, and T3/E3 (45/34 Mbps), which are often too expensive for even large corporate users. Network managers also can bundle WAN services to set up their own ATM networks at native LAN speeds.

IN Intelligent network

Induced uplink An uplink "A" that is created due to the existence of an uplink "B" in the child peer group represented by the node that created uplink "A." Both "A" and "B" share the same upnode, which is higher in the PNNI hierarchy than the peer group in which uplink "A" is seen.

Inside link Synonymous with horizontal link.

Instance ID A subset of an object's attributes that serve to uniquely identify an MIB instance.

Interim interswitch signaling protocol (IISP) Formerly known as PNNI phase 0, this limited, temporary routing scheme requires net managers to establish PVCs between switches from different vendors. Unlike PNNI phase I, which will automatically distribute routing information, IISP relies on static routing tables. Routing is done hop by hop.

Interim local management interface (ILMI) Furnishes a pro-tem basic SNMP management interface for ATM. ILMI requires each user-network interface (UNI) 3.0 end station or ATM network to implement a UNI management entity (UME). The UME functions as an SNMP agent that maintains network and connection information specified in an ILMI MIB (management information base). It responds to requests from SNMP management applications. Standard SNMP management frameworks can use AAL 3/4 or AAL 5 to encapsulate SNMP commands in ATM protocol data units (PDUs).

Interior Denotes that an item (e.g., link, node, or reachable address) is inside of a PNNI routing domain.

Internal reachable address   An address of a destination that is directly attached to the logical node advertising the address.

I/O   Input/output

IOP   Interoperability: The ability of equipment from different manufacturers (or different implementations) to operate together.

IP   Internet protocol: Originally developed by the Department of Defense to support interworking of dissimilar computers across a network. This protocol works in conjunction with TCP and is usually identified as TCP/IP. A connectionless protocol that operates at the network layer (layer 3) of the OSI model.

IPATM   IP over ATM

Ipng   Internet protocol next generation

IPX   Novell Internetwork Packet Exchange: A built-in networking protocol for Novell Netware. It was derived from the Xerox Network System protocol and operates at the network layer of the OSI protocol model.

IS   Intermediate system: A system that provides forwarding functions or relaying functions or both for a specific ATM connection. OAM cells may be generated and received.

ISA   Industry standard architecture

ISDN   Integrated Services Digital Network

ISO   International Standards Organization: An international organization for standardization based in Geneva, Switzerland, that establishes voluntary standards and promotes global trade of 90 member countries.

ITU   International Telecommunications Union

ITU H.222   An ITU-T Study Group 15 standard that addresses the multiplexing of multimedia data on an ATM network.

ITU Q.2100   B-ISDN signaling ATM adaptation layer overview.

ITU Q.2110 B-ISDN adaptation layer—service-specific connection-oriented protocol.

ITU Q.2130 B-ISDN adaptation layer—service-specific connection-oriented function for support of signaling at the UNI.

ITU Q.2931 The signaling standard for ATM to support switched virtual connections. This is based on the signaling standard for ISDN.

ITU Q.931 The signaling standard for ISDN to support SVCs. The basis for the signaling standard developed for frame relay and ATM.

ITU Q.933 The signaling standard for frame relay to support SVCs. This is based on the signaling standard for ISDN.

ITU-T International Telecommunications Union Telecommunications: ITU-T is an international body of member countries whose task is to define recommendations and standards relating to the international telecommunications industry. The fundamental standards for ATM have been defined and published by the ITU-T (previously CCITT).

IUT Implementation under test: The particular portion of equipment that is to be studied for testing. The implementation may include one or more protocols.

IWF Interworking function

IWU Interworking unit

Joining The phase in which the LE client establishes its control connections to the LE server.

JPEG Joint Photographic Experts Group: An ISO Standards group that defines how to compress still pictures.

LAN Local area network: A network designed to move data between stations within a campus.

LANE LAN emulation: The set of services, functional groups, and protocols that provide for the emulation of LANS utilizing ATM as a backbone to allow connectivity among LAN and ATM attached end stations.

LAN emulation network-to-network interface (LNNI)   Enables one vendor's implementation of LAN emulation to work with another's. This spec is essential for building multivendor ATM networks.

LAN emulation user network interface (L-UNI)   Defines   how   legacy LAN applications and protocols work with ATM. L-UNI adapts layer 2 LAN packets to AAL 5 PDUs, which can then be divided into cells. L-UNI uses a client-server architecture to resolve LAN-to-ATM addresses, the most complex aspect of LAN emulation. A LAN emulation client (LEC) resides in each ATM-attached device; a LAN emulation server (LES) and broadcast and unknown server (BUS) reside anywhere on the ATM network. When a legacy LAN end station sends a message across the ATM network to another legacy end station, the LEC requests ATM address and routing information from the LES and BUS, which correlate the MAC-layer LAN address of the destination with the ATM addresses needed to traverse the backbone.

LAN emulation   A way for legacy LAN MAC-layer protocols like Ethernet and token ring, and all higher layer protocols and applications, to access work transparently across an ATM network. LAN emulation retains all Ethernet and token ring drivers and adaptors; no modifications need to be made to Ethernet or token ring end stations.

LAPD   Link access procedure D: A layer 2 protocol defined by CCITT (original name of ITU-T). This protocol reliably transfers blocks of information across a single layer 1 link and supports multiplexing of different connections at layer 2.

LATA   Local access and transport area

Layer entity   An active element within a layer.

Layer function   A part of the activity of the layer entities.

Layer service   A capability of a layer and the layers beneath it that is provided to the upper layer entities at the boundary between that layer and the next higher layer.

Layer user data   Data transferred between corresponding entities on behalf of the upper layer or layer management entities for which they are providing services.

LB Leaky bucket: Leaky bucket is the term used as an analogous description of the algorithm used for conformance checking of cell flows from a user or network. See GCRA, UPC, and NPC. The "leaking hole in the bucket" applies to the sustained rate at which cells can be accommodated, while the "bucket depth" applies to the tolerance to cell bursting over a given time period.

LD LAN destination

LE LAN emulation. Refer to LANE.

Leadership priority The priority with which a logical node wishes to be elected peer group leader of its peer group. Generally, of all nodes in a peer group, the one with the highest leadership priority will be elected as peer group leader.

Leaky bucket algorithm A form of flow control that checks an arriving data stream against the traffic-shaping parameters specified by the sender. Cells arriving at a switch are placed in a "bucket" (memory buffer), which is allowed to fill up but not overflow. The bucket is "leaky" in that it allows cells to flow out to their destinations, allowing more to be added. Incoming cells that would cause the bucket to overflow are considered "nonconforming" (exceeding bandwidth allocations) and are dropped.

Leaky bucket An informal term for the generic cell rate algorithm.

LEC LAN emulation client: The entity in end systems that performs data forwarding, address resolution, and other control functions.

LEC Local exchange carrier: A telephone company affiliate of a Regional Bell Operating Company or an independent telephone company.

LECID LAN emulation client identifier: This identifier, contained in the LAN emulation header, indicates the ID of the ATM host or ATM-LAN bridge. It is unique for every ATM client.

LECS LAN emulation configuration server: This implements the policy-controlled assignment of individual LE clients to different emulated LANs by providing the LES ATM addresses.

LES   LAN emulation server: This implements the control coordination function for the emulated LAN; examples are enabling a LEC to join an ELAN or resolving MAC to ATM addresses.

LE_ARP   LAN emulation address resolution protocol: A message issued by an LE client to solicit the ATM address of another function.

LGN   Logical group node: LGN is a single node that represents the lowest level peer groups in the respective higher level peer group.

LIJP   Leaf initiated joint parameter: Root screening options and information element (IE) instructions carried in SETUP message.

Link aggregation token   Refer to aggregation token.

Link attribute   A link state parameter that is considered individually to determine whether a given link is acceptable and/or desirable for carrying a given connection.

Link connection   A link connection (e.g., at the VP level) is a connection capable of transferring information transparently across a link without adding any overhead, such as cells, for purposes for monitoring. It is delineated by connection points at the boundary of the subnetwork.

Link constraint   A restriction on the use of links for path selection for a specific connection.

Link metric   A link parameter that requires the values of the parameter for all links along a given path to be combined to determine whether the path is acceptable and/or desirable for carrying a given connection.

Link state parameter   Information that captures an aspect or property of a link.

Link   An entity that defines a topological relationship (including available transport capacity) between two nodes in different subnetworks. Multiple links may exist between a pair of subnetworks. Synonymous with logical link.

LIV   Link integrity verification

LLATMI   Lower layer ATM interface

LLC Logical link control

LLC/SNAP Logical link control/subnetwork access protocol

LMI Layer management interface

LNNI LAN emulation network node interface: The interface between two LANE servers (not to be confused with NNI [network node interface], the interface between ATM switches and networks). LANE 1.0 only defines a single server, but multiple servers are needed if LAN emulation is to scale across very large networks and ensure interoperability among LANE service components. LNNI is a part of the LANE 2.0, which is expected to be finished by April 1997.

LNNI LANE NNI: The standardized interface between two LAN servers (LES-LES, BUS-BUS, LECS-LECS, and LECS-LES).

LOC Loss of cell delineation: A condition at the receiver or a maintenance signal transmitted in the PHY overhead indicating that the receiving equipment has lost cell delineation. Used to monitor the performance of the PHY layer.

LOF Loss of frame: A condition at the receiver or a maintenance signal transmitted in the PHY overhead indicating that the receiving equipment has lost frame delineation. This is used to monitor the performance of the PHY layer.

Logical group node A logical node that represents a lower level peer group as a single point for purposes of operating at one level of the PNNI routing hierarchy.

Logical link An abstract representation of the connectivity between two logical nodes. This includes individual physical links, individual virtual path connections, and parallel physical links and/or virtual path connections.

Logical node ID A string of bits that unambiguously identifies a logical node within a routing domain.

Logical node An abstract representation of a peer group or a switching system as a single point.

LOP Loss of pointer: A condition at the receiver or a maintenance signal transmitted in the PHY overhead indicating that the receiving equipment has lost the pointer to the start of cell in the payload. This is used to monitor the performance of the PHY layer.

LOS Loss of signal: A condition at the receiver or a maintenance signal transmitted in the PHY overhead indicating that the receiving equipment has lost the received signal. This is used to monitor the performance of the PHY layer.

LPF Lowpass filter: In an MPEG-2 clock recovery circuit, it is a technique for smoothing or averaging changes to the system clock.

LSAP Link service access point: Logical address of boundary between layer 3 and LLC sublayer 2.

LSB Least significant bit: The lowest order bit in the binary representation of a numerical value.

LSR Leaf setup request: A setup message type used when a leaf node requests connection to an existing point-to-multipoint connection or requests creation of a new multipoint connection.

LT Lower tester: The representation in ISO/IEC 9646 of the means of providing, during test execution, indirect control and observation of the lower service boundary of the IUT using the underlying service provider.

LTE SONET Lite terminating equipment: ATM equipment terminating a communications facility using a SONET Lite transmission convergence (TC) layer. This is usually reserved for end-user or LAN equipment. The SONET Lite TC does not implement some of the maintenance functions used in long-haul networks such as termination of path, line, and section overhead.

LTH Length field

LUNI LANE UNI: The standardized interface between a LE client and a LE Server (LES, LECS and BUS).

M1 Management interface 1: The management of ATM end devices.

M2 Management interface 2: The management of private ATM networks or switches.

M3 Management interface 3: The management of links between public and private networks.

M4 Management interface 4: The management of public ATM networks.

M5 Management interface 5: The management of links between two public networks.

MA Maintenance and adaptation

MAC Media access control: IEEE specifications for the lower half of the data link layer (layer 2) that defines topology dependent access control protocols for IEEE LAN specifications.

MAN Metropolitan area network: A network designed to carry data over an area larger than a campus, such as an entire city and its outlying area.

Managed system An entity that is managed by one or more management systems, which can be either element management systems, subnetwork or network management systems, or any other management system.

Management domain An entity used here to define the scope of naming.

Management system An entity that manages a set of managed systems, which can be either NEs, subnetworks or other management systems.

MARS Multicast address resolution server

MaxCR Maximum cell rate: This is the maximum capacity usable by connections belonging to the specified service category.

Max CTD Maximum cell transfer delay

MBS Maximum burst size: In the signaling message, the burst tolerance (BT) is conveyed through the MBS, which is coded as a number of cells. The BT together with the SCR and the GCRA determine the MBS that may be transmitted at the peak rate and still be in conformance with the GCRA.

MC Multiway communications: A bidirectional multipoint-to-multipoint link that establishes any-to-any communications among all parties at either end of a link. Multiway communications would be appropriate for applications like multipoint videoconferencing and remote learning (when students are geographically dispersed). ATM currently supports bidirectional point-to-point and unidirectional point-to-multipoint (multicast) connections. Although it is technically possible to set up multiway communications, there is no standard yet. At the June 1996, meeting of the ATM Forum, several vendors proposed establishing a formal working group to develop a multiway communications specification. The group would need to address signaling and resource-control protocols along with other issues.

MCDV   Maximum cell delay variance: This is the maximum two-point CDV objective across a link or node for the specified service category.

MCLR   Maximum cell loss ratio: This is the maximum ratio of the number of cells that do not make it across the link or node to the total number of cells arriving at the link or node.

MCR   Minimum cell rate: An ABR service traffic descriptor, in cells/sec, that is the rate at which the source is always allowed to send.

MCTD   Maximum cell transfer delay: This is the sum of the fixed delay component across the link or node and MCDV.

ME   Mapping entity

Mean CTV   Mean cell transfer delay

Metasignaling VCs   The standardized VCs that convey metasignaling information across a user-to-network interface (UNI).

Metasignaling   ATM layer management (LM) process that manages different types of signaling and possibly semipermanent virtual channels (VCs), including the assignment, removal, and checking of VCs.

Metasignaling   The technique employed by the ATM user-to-network interface (UNI) to establish a virtual circuit (VC) that conveys signaling information about switched virtual circuits (SVCs). Metasignaling involves signaling VC assignment, in which the user device and ATM network exchange single-cell messages concerning circuit location; checking, in which the network polls the signaling VC to make sure it's still up; and removal, in which the user device and network exchange cells regarding teardown.

MGF   Moment generating function

MIB   Management information base: A definition of management items for some network component that can be accessed by a network manager. An MIB includes the names of objects it contains and the type of information retained.

MIB attribute   A single piece of configuration, management, or statistical information that pertains to a specific part of the PNNI protocol operation.

MIB instance   An incarnation of an MIB object that applies to a specific part, piece, or aspect of the PNNI protocol's operation.

MIB object   A collection of attributes that can be used to configure, manage, or analyze an aspect of the PNNI protocol's operation.

MID   Message identifier: The message identifier is used to associate ATM cells that carry segments from the same higher layer packet.

MIN   Multistage interconnection networks

MIR   Maximum information rate: Refer to PCR.

MMF   Multimode fiber-optic cable: Fiber-optic cable in which the signal or light propagates in multiple modes or paths. Since these paths may have varying lengths, a transmitted pulse of light may be received at different times and smeared to the point that pulses may interfere with surrounding pulses. This may cause the signal to be difficult or impossible to receive. This pulse dispersion sometimes limits the distance over which a MMF link can operate.

MOS   Media operating system

MPEG   Motion Picture Experts Group: An ISO Standards group dealing with video and audio compression techniques and mechanisms for multiplexing and synchronizing various media streams.

MPEG-2   Motion Picture Experts Group-2

MPOA Multiprotocol over ATM: An effort taking place in the ATM Forum to standardize protocols for the purpose of running multiple network layer protocols over ATM.

MPOA client A device that implements the client side of one or more of the MPOA protocols, (i.e., is an SCP client and/or an RDP client). An MPOA client is either an edge device functional group (EDFG) or a host behavior functional group (HBFG).

MPOA server An MPOA server is any one of an ICFG or RSFG.

MPOA service area The collection of server functions and their clients. A collection of physical devices consisting of an MPOA server plus the set of clients served by that server.

MPOA target A set of protocol address, path attributes, (e.g., internetwork layer QoS, other information derivable from received packet) describing the intended destination and its path attributes that MPOA devices may use as lookup keys.

MRCS Multirate circuit switching

Mrm An ABR service parameter that controls allocation of bandwidth between forward RM-cells, backward RM-cells, and data cells.

MS Metasignaling

MSAP Management service access point

MSB Most significant bit: The highest order bit in the binary representation of a numerical value.

MSN Monitoring cell sequence number

MSP Multistream protocol

MSVC Metasignaling virtual channel

MT Message type: Message type is the field containing the bit flags of an RM-cell. These flags are as follows: DIR = 0 for forward RM-cells = 1 for backward; RM-cells BN = 1 for nonsource-generated (BECN); RM-cells = 0 for source-generated; RM-cells CI = 1 to indicate congestion = 0;

otherwise NI = 1 to indicate no additive increase allowed = 0; otherwise RA—not used for ATM Forum ABR.

MTP Message transfer part: Level 1 through 3 protocols of the SS7 protocol stack. MTP 3 (level 3) is used to support BISUP.

MTU Message transfer unit

Multicast address resolution server (MARS) Comparable to the address resolution protocol (ARP) specified in classical IP over ATM (IETF RFC 1577), MARS resolves IP addresses to a single ATM multicast address that defines an entire group or "cluster" of endpoints. End stations in the cluster use the ATM address of MARS, which maintains point-to-multipoint connections with the members of various multicast groups.

Multicasting The transmit operation of a single PDU by a source interface where the PDU reaches a group of one or more destinations.

MultiPHY Multiphysical port: Part of UTOPIA (universal test and operation physical interface for ATM), MultiPHY defines a way for a single chip or device that implements the ATM physical (PHY) layer to support multiple physical ports—lowering equipment costs by using less silicon for more ports.

Multiplexing A function within a layer that interleaves the information from multiple connections into one connection.

Multipoint access User access in which more than one terminal equipment (TE) is supported by a single network termination.

Multipoint-to-multipoint connection A multipoint-to-multipoint connection is a collection of associated ATM VC or VP links, and their associated nodes, with the following properties: All nodes in the connection, called endpoints, serve as a root node in a point-to-multipoint connection to all of the (N - 1) remaining endpoints.

Multipoint-to-point connection A point-to-multipoint connection may have zero bandwidth from the root node to the leaf nodes, and nonzero return bandwidth from the leaf nodes to the root node. Such a connection is also known as a multipoint-to-point.

Multiprotocol encapsulation over ATM   Allows   high-layer   protocols, such as IP or IPX, to be routed over ATM by enabling an ATM-aware device or application to add a standard protocol identifier to LAN data.

Multiprotocol over ATM (MPOA)   A proposed ATM Forum spec that defines how ATM traffic is routed from one virtual LAN to another. MPOA is key to making LAN emulation, classical IP over ATM, and proprietary virtual LAN schemes interoperate in a multiprotocol environment. Eventually, MOPA will have to deal with conventional routers, distributed ATM edge routers (which shunt LAN traffic across an ATM cloud while also performing conventional routing functions between non-ATM networks), and route servers (which centralize lookup tables on a dedicated network server in a switched LAN).

N-ISDN   Narrowband Integrated Services Digital Network: Services include basic rate interface (2B + D or BRI) and primary rate interface (23B + D or PRI). Supports narrowband speeds at/or below 1.5 Mbps.

Native address   An address that matches one of a given node's summary addresses.

NC-REN   North Carolina Research & Education Network

NCIH   North Carolina Information Highway

NDIS   Network Driver Interface Specification: Refer to 3COM/Microsoft, LAN Manager: Network Driver Interface Specification, October 8, 1990.

NE   Network element: A system that supports at least NEFs and may also support operation system functions/mediation functions. An ATM NE may be realized as either a standalone device or a geographically distributed system. It cannot be further decomposed into managed elements in the context of a given management function.

NEBIOS   Network basic input/output system

NEF   Network element function: A function within an ATM entity that supports the ATM based network transport services, (e.g., multiplexing, cross-connection).

Neighbor node   A node that is directly connected to a particular node via a logical link.

NEL Network element layer: An abstraction of functions related specifically to the technology, vendor, and the network resources or network elements that provide basic communications services.

Network-to-network interface (NNI) Interface between ATM network nodes (switches) defined in the ATM Forum's UNI (user-to-network interface).

NEXT Near end crosstalk: Equipment that must concurrently receive on one wire pair and transmit on another wire pair in the same cable bundle must accommodate NEXT interference. NEXT is the portion of the transmitted signal that leaks into the receive pair. Since at this point on the link the transmitted signal is at maximum and the receive signal

NHRP Next hop resolution protocol

NIC Network interface card

NM Network management entity: The body of software in a switching system that provides the ability to manage the PNNI protocol. NM interacts with the PNNI protocol through the MIB.

NML Network management layer: An abstraction of the functions provided by systems that manage network elements on a collective basis so as to monitor and control the network end to end.

NMS Network management system: An entity that implements functions at the network management layer. It may also include element management layer functions. A network management system may manage one or more other network management systems.

NMS environment A set of NMS that cooperate to manage one or more subnetworks.

NN Neutral network

NNI Network node interface: An interface between ATM switches defined as the interface between two network nodes.

Nodal attribute A nodal state parameter that is considered individually to determine whether a given node is acceptable and/or desirable for carrying a given connection.

Nodal constraint   A restriction on the use of nodes for path selection for a specific connection.

Nodal metric   A nodal parameter that requires the values of the parameter for all nodes along a given path to be combined to determine whether the path is acceptable and/or desirable for carrying a given connection.

Nodal state parameter   Information that captures an aspect or property of a node.

Node   Synonymous with logical node.

NP   Network performance

NPC   Network parameter control: Network parameter control is defined as the set of actions taken by the network to monitor and control traffic from the NNI. Its main purpose is to protect network resources from malicious as well as unintentional misbehavior that can affect the QoS of other already established connections by detecting violations of negotiated parameters and taking appropriate actions. Refer to UPC.

NREN   National Research and Educational Network

NRM   Network resource management

Nrm   An ABR service parameter, Nrm is the maximum number of cells a source may send for each forward RM-cell.

NRT   Nonreal time

NSAP   Network service access point: OSI generic standard for a network address consisting of 20 octets. ATM has specified E.164 for public network addressing and the NSAP address structure for private network addresses.

NSP   Network service provider

NSR   Nonsource routed: Frame forwarding through a mechanism other than source route bridging.

NT Network termination: Network termination represents the termination point of a virtual channel, virtual path, or virtual path/virtual channel at the UNI.

NTSC National Television System Committee: An industry group that defines how television signals are encoded and transmitted in the United States.

NTT Nippon Telegraph and Telephone

Nucleus The interior reference point of a logical node in the PNNI complex node representation.

nx64K This refers to a circuit bandwidth or speed provided by the aggregation of n × 64-Kbps channels (where n= integer 1). The 64K or DS0 channel is the basic rate provided by the T carrier systems.

OA&M Operations, administration, and management

OAM Operations, administration, and maintenance: A group of network management functions that provide network fault indication, performance information, and data and diagnosis functions.

Octet A term for eight (8) bits that is sometimes used interchangeably with "byte" to mean the same thing.

ODI Open Data-Link Interface: This refers to Novell Incorporated, Open Data-Link Interface Developer's Guide, March 20, 1992.

OLI Originating line information

One hop set A set of hosts that are one hop apart in terms of inter-network protocol TTLs (TTL = 0 -on the wire+).

OOF Out of frame: Refer to LOF.

Operations, administration, and maintenance (OAM) flow reference architecture A range of diverse network management functions performed by dedicated ATM cells, including fault and performance management (operations); addressing, data collection, and usage monitoring (administration); and analysis, diagnosis, and repair of network faults (maintenance). OAM cells do not help segmentation and reassembly.

Operations, administration, and maintenance (OAM) flow reference architecture   This reference model also known as the management plane reference architecture, defines the aspects of an ATM point-to-point virtual circuit (VC) that can be monitored and controlled using specialized OAM cells. The reference model divides a VC into five distinct layers, labeled F1 through F5. It also defines the flows of ATM cells through these layers. The F levels are as follows. The F1 level defines the flow of cells at the lowest physical layer of the ATM stack, the SONET (Synchronous Optical Network) section layer (also known as the regeneration section level). A typical transmission path for cells at F1 would be through a SONET repeater in a WAN. The F2 level defines the flow of cells at the SONET line layer (also called the digital section level). An example of an F2 function is the transmission of cells between two lightwave terminal equipment devices in a SONET network. The F3 level partially defines the flow between a virtual path (VP) and a VC. In a large ATM network, a VC typically joins a VP, traverses it, and then splits out into a separate VC again. Traffic is forwarded from the VC to the VP and back to a VC again. Traffic is forwarded from the VC to the VP and back to a VC again via a cell relaying function (CRF). F3 defines the flows between the VC and CRF and between two CRFs. The F4 level completes the definition of the traffic flow between a VP and a VC. F4 describes the transmission of cells from an end station, across a VC, through a CRF, and onto a VP. F4 stops at the second CRF (also known as the VP CRF). The F5 level completes the definition of the traffic flow from one ATM end station to another. The flow goes from VC to CRF to VP and VP CRF, then to VC again, and finally to the destination.

OPCR   Original program clock reference

Optical carrier (OC-n)   Fundamental unit in the SONET (Synchronous Optical Network) hierarchy. OC indicates an optical signal and represents increments of 51.84 Mbps. Thus, OC-1, -3, and -12 equal optical signals of 51, 155, and 622 Mbps.

OSFP   Open shortest first path

OSI   Open Systems Interconnection: A seven (7) layer architecture model for communications systems developed by the ISO for the interconnection of data communications systems. Each layer uses and builds on the services provided by those below it.

OSID   Origination signaling identifier

OSPF Open shortest path first: A link-state routing algorithm that is used to calculate routes based on the number of routers, transmission speed, delays, and route cost.

OUI Organizationally unique identifier: The OUI is a 3-octet field in the IEEE 802.1a defined subnetwork attachment point (SNAP) header identifying an organization that administers the meaning of the following 2-octet protocol identifier (PID) field in the SNAP header. Together, they identify a distinct routed or bridged protocol.

Outlier A node whose exclusion from its containing peer group would significantly improve the accuracy and simplicity of the aggregation of the remainder of the peer group topology.

Outside link A link to an outside node.

Outside node A node that is participating in PNNI routing, but that is not a member of a particular peer group.

PAD Packet assembler and disassembler: A PAD assembles packets of asynchronous data and emits these buffers in a burst to a packet switch network. The PAD also disassembles packets from the network and emits the data to the nonpacket device.

Parent node The logical group node that represents the containing peer group of a specific node at the next higher level of the hierarchy.

Parent peer group The parent peer group of a peer group is the one containing the logical group node representing that peer group. The parent peer group of a node is the one containing the parent node of that node.

Path constraint A bound on the combined value of a topology metric along a path for a specific connection.

Payload type indicator (PTI) field A 3-bit field in the ATM cell header. The first bit indicates which AAL was used to format the data in the payload; the second provides explicit forward congestion indication (EFCI), which alerts the application of possible delays by informing it of congestion behind the cell; the third indicates whether the cell contains data OAM information.

Payload   Information portion of an ATM cell, exclusive of header. ATM cells typically have 48-byte payloads, but size can vary depending upon type of data and AAL.

PBX   Private branch exchange: PBX is the term given to a device that provides private local voice switching and voice-related services within the private network. A PBX could have an ATM API to utilize ATM services (for example, circuit emulation service).

PC   Protocol control: Protocol control is a mechanism that a given application protocol may employ to determine or control the performance and health of the application. For example, protocol liveness may require that protocol control information be sent at some minimum rate; some applications may become intolerable to users if they are unable to send at least some minimum rate. For such applications, the concept of MCR is defined. Refer to MCR.

PCM   Pulse code modulation: An audio encoding algorithm that encodes the amplitude of a repetitive series of audio samples. This encoding algorithm converts analog voice samples into a digital bit stream.

PCO   Point of control and observation: A place (point) within a testing environment where the occurrence of test events is to be controlled and observed as defined by the particular abstract test method used.

PCR   Peak cell rate: The peak cell rate, in cells/sec, is the cell rate that the source may never exceed.

PCR   Program clock reference: A timestamp that is inserted by the MPEG-2 encoder into the transport stream to aid the decoder in recovering and tracking the encoder clock.

PCR   Peak cell rate: The maximum rate at which cells can be transmitted across a virtual circuit, specified in cells per second and defined by the interval between the transmission of the last bit of one cell and the first bit of the next.

PCS   Personal communications services

PCVS   Point-to-point switched virtual connections

PD   Packetization delay

PDH Plesiochronous digital hierarchy: PDH (plesiochronous means nearly synchronous), was developed to carry digitized voice over twisted pair cabling more efficiently. This evolved into the North American, European, and Japanese digital hierarchies where only a discrete set of fixed rates is available, namely, nxDS0 (DS0 is a 64-Kbps rate) and then the next levels in the respective multiplex hierarchies.

PDU Packet data unit

PDU Protocol data unit: A PDU is a message of a given protocol comprising payload and protocol-specific control information, typically contained in a header. PDUs pass over the protocol interfaces that exist between the layers of protocols (per OSI model).

Peer entities Entities within the same layer.

Peer group A set of logical nodes that are grouped for purposes of creating a routing hierarchy. PTSEs are exchanged among all members of the group.

Peer group identifier A string of bits that is used to unambiguously identify a peer group.

Peer group leader A node that has been elected to perform some of the functions associated with a logical group node.

Peer group level The number of significant bits in the peer group identifier of a particular peer group.

Peer node A node that is a member of the same peer group as a given node.

Permanent virtual circuit (PVC) A virtual link with fixed endpoints that are defined by the network manager. A single virtual path may support multiple PVCs.

PES Packetized elementary stream: In MPEG-2, after the media stream has been digitized and compressed, it is formatted into packets before it is multiplexed into either a program stream or transport stream.

PG Peer group: A set of logical nodes that are grouped for purposes of creating a routing hierarchy. PTSEs are exchanged among all members of the group.

PGL   Peer group leader: A single real physical system that has been elected to perform some of the functions associated with a logical group node.

PHY   OSI physical layer: The physical layer provides for transmission of cells over a physical medium connecting two ATM devices. This physical layer is comprised of two sublayers: the PMD physical medium dependent sublayer, and the TC transmission convergence sublayer. Refer to PMD and TC.

Physical layer (PHY) connection   An association established by the PHY between two or more ATM entities. A PHY connection consists of the concatenation of PHY links in order to provide an end-to-end transfer capability to PHY SAPs.

Physical layer (PHY)   The bottom layer of the ATM protocol stack, which defines the interface between ATM traffic and the physical media. The PHY consists of two sublayers: the physical medium-dependent (PMD) sublayer and the transmission convergence (TC) sublayer.

Physical layer convergence protocol (PLCP)   A   protocol   specified within the TC sublayer that defines how cells are formatted within a data stream for a particular transmission facility, such as T1, T3, or OC-n.

Physical link   A real link that attaches two switching systems.

Physical medium-dependent (PMD) sublayer   Defines the actual speed at which ATM traffic can be transmitted across a given physical medium. The ATM Forum has approved three SONET interfaces for UNI: STS-1 at 51.84 Mbps, STS-3c at 155.52 Mbps, and STS-12c at 622.08 Mbps, as well as DS-1 (T1) at 1.544 Mbps, E1 at 2.048 Mbps, E3 at 34.368 Mbps, and DS-3 (T3) at 44.73 Mbps. The Forum also has adopted a number of specifications for LAN environments, including a 100-Mbps interface using FDDI encoding, a 155-Mbps interface using category 5 UTP (unshielded twisted pair), and a 51-Mbps interface using category 3 UTP.

PICS   Protocol implementation conformance statement: A statement made by the supplier of an implementation or system stating which capabilities have been implemented for a given protocol.

PID   Protocol identification. Refer to OUI.

PIXIT Protocol implementation extra information for testing: A statement made by a supplier or implementor of an IUT that contains information about the IUT and its testing environment that will enable a test laboratory to run an appropriate test suite against the IUT.

PL Physical layer

Plastic fiber optics An optical fiber where the core transmission media is plastic, in contrast to glass or silica cores. Proposed plastic fibers generally have larger attenuation and dispersion than glass fiber but may have applications where the distance is limited. Plastic systems may also offer lower cost connectors that may be installed with simple tools and a limited amount of training.

PLCP Physical layer convergence protocol: The PLCP is defined by the IEEE 802.6. It is used for DS3 transmission of ATM. ATM cells are encapsulated in a 125-ms frame defined by the PLCP, which is defined inside the DS3 M-frame.

PLL Phase lock loop: Phase lock loop is a mechanism whereby timing information is transferred within a data stream and the receiver derives the signal element timing by locking its local clock source to the received timing information.

PM Physical medium: Physical medium refers to the actual physical interfaces. Several interfaces are defined, including STS-1, STS-3c, STS-12c, STM-1, STM-4, DS1, E1, DS2, E3, DS3, E4, FDDI-based, fiber channel-based, and STP. These range in speeds from 1.544 Mbps through 622.08 Mbps.

PMD Physical media dependent: This sublayer defines the parameters at the lowest level, such as speed of the bits on the media.

PNI Permit next increase: An ABR service parameter, PNI is a flag controlling the increase of ACR upon reception of the next backward RM-cell. PNI = 0 inhibits increase. The range is 0 or 1.

PNNI Private network-network interface: A routing information protocol that enables extremely scalable, full function, dynamic multivendor ATM switches to be integrated in the same network.

PNNI crankback A feature of the private network-to-network interface (PNNI) that enables a blocked connection on an ATM network to return to an earlier point in its path and select another route. Crankback allows intermediate switches on the path to exploit new information regarding network conditions and prevents unnecessary retransmissions.

PNNI protocol entity The body of software in a switching system that executes the PNNI protocol and provides the routing service.

PNNI routing control channel VCCs used for the exchange of PNNI routing protocol messages.

PNNI routing domain A group of topologically contiguous systems that are running one instance of PNNI routing.

PNNI routing hierarchy The hierarchy of peer groups used for PNNI routing.

PNNI topology state element A collection of PNNI information that is flooded among all logical nodes within a peer group.

PNNI topology state packet A type of PNNI routing packet that is used for flooding PTSEs among logical nodes within a peer group.

POH Path overhead: A maintenance channel transmitted in the SONET overhead following the path from the beginning multiplexer to the ending demultiplexer. This is not implemented in SONET Lite.

POI Path overhead indicator

Point-to-multipoint connection A point-to-multipoint connection is a collection of associated ATM VC or VP links, with associated endpoint nodes, with the following properties. (1) One ATM link, called the root link, serves as the root in a simple tree topology. When the root node sends information, all of the remaining nodes on the connection, called leaf nodes, receive copies of the information. (2) Each of the leaf nodes on the connection can send information directly to the root node. The root node cannot distinguish which leaf is sending information without additional (higher layer) information. (See note below for UNI 4.0 support). (3) The leaf nodes cannot communicate directly to each other with this connection type.
Note: UNI 4.0 does not support traffic sent from a leaf to the root.

Point-to-point connection A connection with only two endpoints.

POP Point of presence

Port identifier The identifier assigned by a logical node to represent the point of attachment of a link to that node.

PRI Primary rate interface: An ISDN standard for provisioning of 1.544-Mbps (DS1) ISDN services. The standard supports 23 "B" channels of 64 Kbps each and one "D" channel of 64 Kbps.

Primitive An abstract, implementation-independent interaction between a layer service user and a layer service provider.

Private ATM address A 20-byte address used to identify an ATM connection termination point.

Private network-to-network interface (PNNI) A routing information protocol that allows different vendors' ATM switches to be integrated in the same network. PNNI automatically and dynamically distributes routing information, enabling any switch to determine a path to any other switch.

Protocol control information Information exchanged between corresponding entities, using a lower layer connection, to coordinate their joint operation.

Protocol data unit (PDU) A discrete piece of information (such as a packet or frame) in the appropriate format to be segmented and encapsulated in the payload of an ATM cell.

Protocol A set of rules and formats (semantic and syntactic) that determines the communication behavior of layer entities in the performance of the layer functions.

PSTN Public switched telephone network

PT Payload type: Payload type is a 3-bit field in the ATM cell header that discriminates between a cell carrying management information or · one that is carrying user information.

PTI Payload type indicator: Payload type indicator is the payload type field value distinguishing the various management cells and user cells. Example: resource management cell has PTI = 110, end-to-end OAM F5 flow cell has PTI = 101.

PTMPT   Point-to-multipoint: A main source to many destination connections.

PTS   Presentation timestamp: A timestamp that is inserted by the MPEG-2 encoder into the packetized elementary stream to allow the decoder to synchronize different elementary streams (i.e., lip sync).

PTSE   PNNI topology state element: A collection of PNNI information that is flooded among all logical nodes within a peer group.

PTSP   PNNI topology state packet: A type of PNNI routing packet that is used for flooding PTSEs among logical nodes within a peer group.

PVC   Permanent virtual circuit: This is a link with static route defined in advance, usually by manual setup.

PVCC   Permanent virtual channel connection: A virtual channel connection (VCC) is an ATM connection where switching is performed on the VPI/VCI fields of each cell. A permanent VCC is one that is provisioned through some network management function and left up indefinitely.

PVPC   Permanent virtual path connection: A virtual path connection (VPC) is an ATM connection where switching is performed on the VPI field only of each cell. A permanent VPC is one that is provisioned through some network management function and left up indefinitely.

QD   Queuing delay: Queuing delay refers to the delay imposed on a cell by its having to be buffered because of unavailability of resources to pass the cell onto the next network function or element. This buffering could be a result of oversubscription of a physical link or due to a connection of higher priority or tighter service constraints getting the resource of the physical link.

QoS   Quality of service: Quality of service is defined on an end-to-end basis in terms of the following attributes of the end-to-end ATM connection: cell loss ratio; cell transfer delay; cell delay variation.

QPSX   Queue packet and synchronous circuit exchange

Quality of service classes   Five broad categories outlined by the ATM Forum's UNI 3.0; implementation details precise characteristics are to be determined in the future. Class 1 specifies performance requirements and indicates that ATM's quality of service should be comparable with the service offered by standard digital connections. Class 2 specifies necessary service levels for packetized video and voice. Class 3 defines requirements for interoperability with other connection-oriented protocols, particularly frame relay. Class 4 specifies interoperability requirements for connectionless protocols, including IP, IPX, and SMDS. Class 5 is effectively a "best effort" attempt at delivery; it is intended for applications that do not require a particular class of service.

RACE Research for Advanced Communications in Europe

RAI Remote alarm indication

RBOC Regional Bell Operating Company: Seven companies formed to manage the local exchanges originally owned by AT&T. These companies were created as a result of an agreement between AT&T and the United States Department of Justice.

RC Routing control

RD Route descriptor

RD Routing domain: A group of topologically contiguous systems that are running one instance of routing.

RDF Rate decrease factor: An ABR service parameter, RDF controls the decrease in the cell transmission rate. RDF is a power of 2 from 1/32,768 to 1.

RDI Remote defect identification

RDI Remote defect indication

Registration The address registration function is the mechanism by which clients provide address information to the LAN emulation server.

REL Release message

Relaying A function of a layer by means of which a layer entity receives data from a corresponding entity and transmits it to another corresponding entity.

RFC Request for comment: The development of TCP/IP standards, procedures, and specifications is done via this mechanism. RFCs are documents that progress through several development stages, under the control of IETF, until they are finalized or discarded.

RFC1695 Definitions of managed objects for ATM management or AToM MIB.

RFI Radio frequency interface: Refer to EMI.

RI Routing information

RIF Rate increase factor: This controls the amount by which the cell transmission rate may increase upon receipt of an RM-cell. The additive increase rate AIR = PCR * RIF. RIF is a power of 2, ranging from 1/32,768 to 1.

RII Routing information indicator

RIP Routing information protocol

RISC Reduced instruction set computing: A computer processing technology in which a microprocessor understands a few simple instructions thereby providing fast, predictable instruction flow.

RLC Release complete

RM Resource management

RM-Cell Resource management cell: Information about the state of the network like bandwidth availability, state of congestion, and impending congestion, is conveyed to the source through special control cells called resource management cells (RM-cells).

RM Resource management: Resource management is the management of critical resources in an ATM network. Two critical resources are buffer space and trunk bandwidth. Provisioning may be used to allocate network resources in order to separate traffic flows according to service characteristics. VPCs play a key role in resource management. By reserving capacity on VPCs, the processing required to establish individual VCCs is reduced. Refer to RM-cell.

RO Read-only: Attributes that are read-only cannot be written by network management. Only the PNNI protocol entity may change the value of a read-only attribute. Network management entities are restricted to only reading such read-only attributes. Read-only attributes are typically for statistical information, including reporting result of actions taken by autoconfiguration.

ROLC Routing over large clouds

Route server A physical device that runs one or more network layer routing protocols and that uses a route query protocol in order to provide network layer routing forwarding descriptions to clients.

Router A physical device that is capable of forwarding packets based on network layer information and that also participates in running one or more network layer routing protocols.

Routing computation The process of applying a mathematical algorithm to a topology database to compute routes. There are many types of routing computations that may be used. The Dijkstra algorithm is one particular example of a possible routing computation.

Routing constraint A generic term that refers to either a topology constraint or a path constraint.

Routing protocol A general term indicating a protocol run between routers and/or route servers in order to exchange information used to allow computation of routes. The result of the routing computation will be one or more forwarding descriptions.

RS Remote single layer (test method): An abstract test method in which the upper tester is within the system under test and there is a point of control and observation at the upper service boundary of the implementation under test (IUT) for testing one protocol layer. Test events are specified in terms of the abstract service primitives (ASP) and/or protocol data units at the lower tester PCO.

RSE Remote single-layer embedded (test method): An abstract test method in which the upper tester is within the system under test and there is a point of control and observation at the upper service boundary of the implementation under test (IUT) for testing a protocol layer or sublayer which is part of a multiprotocol IUT.

RSFG   Route server functional group: The group of functions performed to provide internetworking level functions in an MPOA system. This includes running conventional interworking routing protocols and providing inter-IASG destination resolution.

RSVP (protocol)   Resource reservation protocol

RT   Real time

RT   Routing type

RTS   Residual timestamp

RW   Read-write: Attributes that are read-write cannot be written by the PNNI protocol entity. Only the network management entity may change the value of a read-write attribute. The PNNI protocol entity is restricted to only reading such read-write attributes. Read-write attributes are typically used to provide the ability for network management to configure, control, and manage a PNNI protocol entity's behavior.

SA   Source MAC address: A 6-octet value uniquely identifying an endpoint and that is sent in an IEEE LAN frame header to indicate source of frame.

SA   Source address: The address from which the message or data originated.

SAA   Service aspects and applications

SAAL   Signaling ATM adaptation layer: This resides between the ATM layer and the Q.2931 function. The SAAL provides reliable transport of Q.2931 messages between Q.2931 entities (e.g., ATM switch and host) over the ATM layer; two sublayers: common part and service-specific part.

SAP   Service access point: A SAP is used for the following purposes. (1) When the application initiates an outgoing call to a remote ATM device, a destination_SAP specifies the ATM address of the remote device plus further addressing that identifies the target software entity within the remote device. (2) When the application prepares to respond to incoming calls from remote ATM devices, a local_SAP specifies the ATM address of the device housing the application plus further addressing that identifies the application within the local device. There are several groups of SAPs that are specified as valid for native ATM services.

SAR Segmentation and reassembly: Method of breaking up arbitrarily sized packets.

SARPVP SONET ATM ring point-to-point virtual path

SCCP Signaling connection and control part: An SS7 protocol that provides additional functions to the message transfer part (MTP). It typically supports transaction capabilities application part (TCAP).

Scope A scope defines the level of advertisement for an address. The level is a level of a peer group in the PNNI routing hierarchy.

SCP Service control point: A computer and database system that executes service logic programs to provide customer services through a switching system. Messages are exchanged with the SSP through the SS7 network.

SCR Sustainable cell rate: The SCR is an upper bound on the conforming average rate of an ATM connection over time scales that are long relative to those for which the PCR is defined. Enforcement of this bound by the UPC could allow the network to allocate sufficient resources, but less than those based on the PCR, and still ensure that the performance objectives (e.g., for cell loss ratio) can be achieved.

SDH Synchronous digital hierarchy: The ITU-TSS International standard for transmitting information over optical fiber.

SDT (structured data transfer) mode SDT is a real-time transfer mode that uses an 8-byte pointer field in each CBR cell to indicate a sequence number. When cell timing varies to some degree and cells arrive at the destination out of sequence, the destination station can use the sequence number to put cells into their proper order. If cell times vary too greatly or too much resequencing is required, however, cells will be dropped. In this event, the destination station can use sequence number information to alert the source that there is a problem, but it does not have enough data to tell the source which cells to resend.

SDT Structured data transfer: An AAL1 data transfer mode in which data is structured into blocks, which are then segmented into cells for transfer.

SDU   Service data unit: A unit of interface information whose identity is preserved from one end of a layer connection to the other.

SE   Switching element: Switching element refers to the device or network node that performs ATM switching functions based on the VPI or VPI/VCI pair.

SEAL   Simple and efficient adaptation layer: An earlier name for AAL5.

Segment   A single ATM link or group of interconnected ATM links of an ATM connection.

Segmentation and reassembly (SAR) sublayer   Converts PDUs into appropriate lengths and formats them to fit the payload of an ATM cell. At the destination end station, SAR extracts the payloads from the cells and converts them back into PDUs, which can be used by applications higher up the protocol stack.

Segmentation and reassembly protocol data unit (SAR-PDU)
Information that has passed through SAR and been loaded into ATM cells that is ready to be forwarded to the TC sublayer of the ATM physical layer for actual transmission.

SEL   Selector: A subfield carried in SETUP message part of ATM endpoint address domain-specific part (DSP) defined by ISO 10589; not used for ATM network routing, used by ATM end systems only.

Semipermanent connection   A connection established via a service order or via network management.

Service-specific convergence sublayer (SSCS)   The portion of the convergence sublayer that is dependent upon the type of traffic that is being converted, such as the frame relay service-specific convergence sublayer (FR-SSCS) or the switched multimegabit data service service-specific convergence sublayer (SMDS-SSCS).

SES   Severely errored seconds: A unit used to specify the error performance of T carrier systems. This indicates a second containing 10 or more errors, usually expressed as SES per hour, day, or week. This method gives a better indication of the distribution of bit errors than a simple bit error rate (BER). Refer also to EFS.

SES   Source end station: An ATM termination point, which is the source of ATM messages of a connection and is used as a reference point for ABR services. Refer to DES.

SF   Superframe: A DS1 framing format in which 24 DS0 timeslots plus a coded framing bit are organized into a frame that is repeated 12 times to form the superframe.

SGM   Segmentation message

Shaping Descriptor   N ordered pairs of GCRA parameters (I, L) used to define the negotiated traffic shape of a connection.

SID   Signaling identifier

Signaling   The standard process, based on CCITT Q.93B, used to establish ATM point-to-point, point-to-multipoint, and multipoint-to-multipoint connections.

Simple and efficient adaptation layer (SEAL)   See   ATM   adaptation layer (AAL 5).

SIPP   SMDS interface protocol: Protocol where layer 2 is based on ATM, AAL, and DQDB. Layer 1 is DS1 and DS3.

SIR   Sustained information rate

SMC   Sleep mode connection

SMDS   Switched multimegabit data services: A connectionless service used to connect LANs, MANs, and WANs to exchange data.

SMF   Single-mode fiber: Fiber-optic cable in which the signal or light propagates in a single mode or path. Since all light follows the same path or travels the same distance, a transmitted pulse is not dispersed and does not interfere with adjacent pulses. SMF fibers can support longer distances and are limited mainly by the amount of attenuation. Refer to MMF.

SN   Sequence number: SN is a 4-octet field in a resource management cell defined by the ITU-T in Recommendation I.371 to sequence such cells. It is not used for ATM Forum ABR. An ATM switch will either preserve this field or set it in accordance with I.371.

SNA Systems Network Architecture: IBM's seven layer, vendor specific architecture for data communications.

SNAP Subnetwork access protocol

SNC Subnetwork connection: In the context of ATM, an entity that passes ATM cells transparently, (i.e., without adding any overhead). An SNC may be either a standalone SNC or a concatenation of SNCs and link connections.

SN cell Sequence number cell: A cell sent periodically on each link of an AIMUX to indicate how many cells have been transmitted since the previous SN cell. These cells are used to verify the sequence of payload cells reassembled at the receiver.

SNMP Simple Network Management Protocol: Originally designed for the Department of Defense network to support TCP/IP network management. It has been widely implemented to support the management of a broad range of network products and functions. SNMP is the IETF standard management protocol for TCP/IP networks.

SOH Section overhead

SONET Synchronous Optical Network: An ANSI standard for transmitting information over optical fiber. This standard is used or accepted in the United States and Canada and is a variation of the SDH international standard.

Source route As used in this document, a hierarchically complete source route.

Source traffic A set of traffic parameters belonging to the ATM traffic descriptor used during the connection setup to capture the intrinsic traffic characteristics of the connection requested by the source.

SPANS Simple Protocol for ATM Network Signaling

SPE SONET synchronous payload envelope

Split system A switching system that implements the functions of more than one logical node.

SPID Service protocol identifier

SPTS Single program transport stream: An MPEG-2 transport stream that consists of only one program.

SR Source routing: A bridged method whereby the source at a data exchange determines the route that subsequent frames will use.

SRF Specifically routed frame: A source routing bridging frame that uses a specific route between the source and destination.

SRT Source routing transparent: An IETF bridging standard combining transparent bridging and source route bridging.

SRTS (synchronous residual time stamp) mode SRTS requires that both the origin and destination end stations in an ATM VC have extremely accurate clock frequencies synchronized with the ATM network's reference clock. When an SRTS connection is established, the signaling end station encodes a residual timestamp (RTS) from the network frequency clock and sends it to the destination end station. The destination station, factoring in transmission time, uses the RTS to synchronize its timing with that of the source. Once the two end stations are synchronized, presuming their clocking mechanisms are accurate, the source can transmit to the end station in real time. To maintain accuracy, the signaling end station recomputes the RTS once every eight cell times.

SRTS Synchronous residual timestamp: A clock-recovery technique in which difference signals between source timing and a network reference timing signal are transmitted to allow reconstruction of the source timing at the destination.

SS7 Signal System Number 7: A family of signaling protocols originating from narrowband telephony. They are used to set up, manage, and tear down connections as well as to exchange non-connection-associated information. Refer to BISUP, MTP, SCCP, and TCAP.

SSCF Service-specific coordination function: SSCF is a function defined in Q.2130, B-ISDN Signaling ATM Adaptation Layer-Service Specific

Coordination Function for Support of Signaling at the User-to- Network Interface.

SSCOP Service specific connection-oriented protocol: An adaptation layer protocol defined in ITU-T Specification Q.2110.

SSCS Service-specific convergence sublayer: The portion of the convergence sublayer that is dependent upon the type of traffic that is being converted.

ST Segment type

STC System time clock: The master clock in an MPEG-2 encoder or decoder system.

STE Spanning tree explorer: A source route bridging frame that uses the spanning tree algorithm in determining a route.

STE SONET section terminating equipment: SONET equipment that terminates a section of a link between a transmitter and repeater, repeater and repeater, or repeater and receiver. This is usually implemented in wide area facilities and not implemented by SONET Lite.

STM Synchronous transfer module: STM is a basic building block used for a synchronous multiplexing hierarchy defined by the CCITT/ITU-T. STM-1 operates at a rate of 155.52 Mbps (same as STS-3).

STM-1 Synchronous transport module 1: SDH standard for transmission over OC-3 optical fiber at 155.52 Mbps.

STM-n Synchronous transport module n: (where n is an integer) SDH standards for transmission over optical fiber (OC-'n × 3) by multiplexing n STM-1 frames, (e.g., STM-4 at 622.08 Mbps and STM-16 at 2.488 Gbps).

STM-nc Synchronous transport module n concatenated: (where n is an integer) SDH standards for transmission over optical fiber (OC-'n × 3) by multiplexing n STM-1 frames, (e.g., STM-4 at 622.08 Mbps and STM-16 at 2.488 Gbps, but treating the information fields as a single concatenated payload).

STP Shielded twisted pair: A cable containing one or more twisted pair wires with each pair having a shield of foil wrap.

STP Signaling transfer point: A high-speed, reliable, special-purpose packet switch for signaling messages in the SS7 network.

STS Synchronous timestamps

STS-1 Synchronous transport signal 1: SONET standard for transmission over OC-1 optical fiber at 51.84 Mbps.

STS-3c Synchronous transport system-level 3 concatenated

STS-n Synchronous transport signal n: (where n is an integer) SONET standards for transmission over OC-n optical fiber by multiplexing n STS-1 frames, (e.g., STS-3 at 155.52 Mbps STS-12 at 622.08 Mbps, and STS-48 at 2.488 Gbps).

STS-nc Synchronous transport signal n concatenated: (where n is an integer) SONET standards for transmission over OC-n optical fiber by multiplexing n STS-1 frames, (e.g., STS-3 at 155.52 Mbps STS-12 at 622.08 Mbps, and STS-48 at 2.488 Gbps, but treating the information fields as a single concatenated payload).

Sublayer A logical subdivision of a layer.

Subnet The use of the term subnet to mean a LAN technology is an historical use and is not specific enough in the MPOA work. Refer to internetwork address subgroup, direct set, host apparent address subgroup, and one hop set for more specific definitions.

Subnetwork A collection of managed entities grouped together from a connectivity perspective, according to their ability to transport ATM cells.

subNMS Subnetwork Management System: A network management system that is managing one or more subnetworks and that is managed by one or more network management systems.

Summary address An address prefix that tells a node how to summarize reachability information.

Sustainable cell rate (SCR) Maximum throughput bursty traffic can achieve within a given virtual circuit without risking cell loss.

SUT System under test: The real open system in which the implementation under test (IUT) resides.

SVC Switched virtual circuit: A connection established via signaling. The user defines the endpoints when the call is initiated.

SVCC Switched virtual channel connection: A switched VCC is one which is established and taken down dynamically through control signaling. A virtual channel connection (VCC) is an ATM connection where switching is performed on the VPI/VCI fields of each cell.

SVCI Switched virtual circuit identifier

SVE SAP vector element: The SAP address may be expressed as a vector, (ATM_addr, ATM_selector, BLLI_id2, BLLI_id3, BHLI_id), where: ATM_addr corresponds to the 19 most significant octets of a device's 20-octet ATM address (private ATM address structure) or the entire E.164 address (E.164 address structure) ATM_selector corresponds to the least significant octet of a device's 20-octet ATM address (private ATM address structure only) BLLI_id2 corresponds to an octet in the Q.2931 BLLI information element that identifies a layer 2 protocol. BLLI_id3 corresponds to a set of octets in the Q.2931 BLLI information element that identify a layer 3 protocol. BHLI_id corresponds to a set of octets in the Q.2931 BHLI information element that identify an application (or session layer protocol of an application). Each element of the SAP vector is called a SAP vector element, or SVE. Each SVE consists of a tag, length, and value field.

SVP Switched virtual path

SVPC Switched virtual path connection: A switched virtual path connection is one that is established and taken down dynamically through control signaling. A virtual path connection (VPC) is an ATM connection where switching is performed on the VPI field only of each cell.

SWG Subworking group

Switched connection A connection established via signaling.

Switched virtual circuit (SVC) A virtual link, with variable endpoints, established through an ATM network. With an SVC, the user defines the endpoints when the call is initiated; with a PVC, the endpoints are predefined by the network manager. A single virtual path may support multiple SVCs.

Switching system   A set of one or more systems that act together and appear as a single switch for the purposes of PNNI routing.

Symmetric connection   A connection with the same bandwidth value specified for both directions.

Synchronous Digital Hierarchy (SDH)   International form of SONET. SDH is built on blocks of 155.52 Mbps; SONET, 51.84 Mbps.

Synchronous Optical Network (SONET)   An international suite of standards for transmittal digital information over optical interfaces. "Synchronous" indicates that all component portions of the SONET signal can be tied to a single reference clock.

Synchronous transfer mode (STM)   B-ISDN communications method that transmits a group of different data streams synchronized to a single reference clock. All data receives the same amount of bandwidth. STM is the standard method carriers use to assign timeslots or channels within a T1/E1 leased line.

Synchronous transfer module (STM-n)   Basic unit of SDH (Synchronous Digital Hierarchy), defined in increments of 155.52 Mbps, with n representing multiples of that rate. The most common values of n are 1, 2, and 4.

Synchronous transfer signal (STS-n)   Basic unit of SONET defined in increments of 51.84 Mbps, with n representing multiples of that rate. The most common values of n are 1, 3, and 12. STS uses an electrical rather than an optical signal.

T1   A digital transmission service with a basic data rate of 1.544 Mbps.

T1E1   An ANSI standards subcommittee dealing with network interfaces.

T1M1   An ANSI standards subcommittee dealing with internetwork operations, administration, and maintenance.

T1Q1   An ANSI standards subcommittee dealing with performance.

T1S1 An ANSI standards subcommittee dealing with services, architecture, and signaling.

T1X1 An ANSI standards subcommittee dealing with digital hierarchy and synchronization.

T3 A digital transmission service with a basic data rate of 44.736 Mbps for transport of 28 T1 circuits.

TAXI Transparent asynchronous transmitter/receiver interface

TB Transparent bridging: An IETF bridging standard where bridge behavior is transparent to the data traffic. To avoid ambiguous routes or loops, a spanning tree algorithm is utilized.

TBE Transient buffer exposure: This is a negotiated number of cells that the network would like to limit the source to sending during startup periods, before the first RM-cell returns.

TC Transaction capabilities: TCAP (see below) plus supporting presentation, session and transport protocol layers.

TC Transmission convergence: The TC sublayer transforms the flow of cells into a steady flow of bits and bytes for transmission over the physical medium. On transmit, the TC sublayer maps the cells to the frame format, generates the header error check (HEC), and sends idle cells when the ATM layer has none to send. On reception, the TC sublayer delineates individual cells in the received bit stream and uses the HEC to detect and correct received errors.

TCAP Transaction capabilities applications part: A connectionless SS7 protocol for the exchange of information outside the context of a call or connection. It typically runs over SCCP and MTP 3.

TCI Test cell input

TCO Test cell output

TCP Transmission control protocol: Originally developed by the Department of Defense to support interworking of dissimilar computers across a network. A protocol that provides end-to-end, connection-oriented, reliable transport layer (layer 4) functions over IP controlled networks. TCP performs the following functions: flow control between two systems, acknowledgments of packets received, and end-to-end sequencing of packets.

TCP Test coordination procedure: A set of rules to coordinate the test process between the lower tester and the upper tester. The purpose is to enable the lower tester to control the operation of the upper tester. These procedures may or may not be specified in an abstract test suite.

TCP/IP Transmission control program/Internet protocol

TCR Tagged cell rate: An ABR service parameter, TCR limits the rate at which a source may send out-of-rate forward RM-cells. TCR is a constant fixed at 10 cells/second.

TCS Transmission convergence sublayer: This is part of the ATM physical layer that defines how cells will be transmitted by the actual physical layer.

TDF An ABR service parameter, TDF controls the decrease in ACR associated with TOF. TDF is signaled as TDFF, where TDF = TDFF/RDF times the smallest power of 2 greater or equal to PCR. TDF is in units of 1/seconds.

TDFF Refer to TDF. TDFF is either zero or a power of 2 in the range 1/64 to 1 in units of 1/cells.

TDJ Transfer delay jitter

TDM Time division multiplexing: A method in which a transmission facility is multiplexed among a number of channels by allocating the facility to the channels on the basis of timeslots.

TE Terminal equipment: Terminal equipment represents the endpoint of ATM connection(s) and termination of the various protocols within the connection(s).

TLV Type/length/value: A coding methodology that provides a flexible and extensible means of coding parameters within a frame. Type indicates parameter type. Length indicates parameter's value length. Value indicates the actual parameter value.

TM   Traffic management: Traffic management is the aspects of the traffic control and congestion control procedures for ATM. ATM layer traffic control refers to the set of actions taken by the network to avoid congestion conditions. ATM layer congestion control refers to the set of actions taken by the network to minimize the intensity, spread, and duration of congestion. The following functions form a framework for managing and controlling traffic and congestion in ATM networks and may be used in appropriate combinations: connection admission control, feedback control, usage parameter control, priority control, traffic shaping, network resource management, frame discard, and ABR flow control.

TMP   Test management protocol: A protocol that is used in the test coordination procedures for a particular test suite.

TM SWG   Traffic management subworking group

TMP   Test management protocol

TNS   Transit network selection: A signaling element that identifies a public carrier to which a connection setup should be routed.

TOF   Timeout factor: An ABR service parameter, TOF controls the maximum time permitted between sending forward RM-cells before a rate decrease is required. It is signaled as TOFF where TOF = TOFF + 1. TOFF is a power of 2 in the range: 1/8 to 4,096.

TOFF   Timeout factor: Refer to TOF.

Topology aggregation   The process of summarizing and compressing topology information at a hierarchical level to be advertised at the level above.

Topology attribute   A generic term that refers to either a link attribute or a nodal attribute.

Topology constraint   A topology constraint is a generic term that refers to either a link constraint or a nodal constraint.

Topology database   The database that describes the topology of the entire PNNI routing domain as seen by a node.

Topology metric A generic term that refers to either a link metric or a nodal metric.

Topology state parameter A generic term that refers to either a link parameter or a nodal parameter.

TP4 Transport protocol 4

TPCC Third-party call control: A connection setup and management function that is executed from a third party that is not involved in the data flow.

TP-MIC Twisted pair media interface connector: This refers to the connector jack at the end-user or network equipment that receives the twisted pair plug.

TR Token ring

Traffic management (TM) See congestion control.

Traffic shaping Allows the sender to specify the throughput and priority of information entering the ATM network and to monitor its progress to ascertain if service levels are met.

Trail An entity that transfers information provided by a client layer network between access points in a server layer network. The transported information is monitored at the termination points.

Trailer Protocol control information located at the end of a PDU.

Trailing packet discard (TPD) An intelligent packet discard mechanism used with ATM adaptation layer (AAL) 5 applications that permits an ATM switch to hold down on retransmissions. When one or more of the cells in an AAL 5 frame is lost, the network marks subsequent cells so that a switch further down the line knows it can dump them, thus conserving bandwidth.

Transit delay The time difference between the instant at which the first bit of a PDU crosses one designated boundary and the instant at which the last bit of the same PDU crosses a second designated boundary.

Transmission convergence (TC) sublayer   Part of the ATM physical layer, it defines a protocol for preparing cells for transmission across the physical media defined by the physical media-dependent (PMD) sublayer. The function of the TC sublayer differs according to physical medium.

Trm   An ABR service parameter that provides an upper bound on the time between forward RM-cells for an active source. It is 100 times a power of two with a range of 100 * 2 - 7 to 100 * 20.

TS   Timeslot

TS   Timestamp: Timestamping is used on OAM cells to compare time of entry of cell to time of exit of cell to be used to determine the cell transfer delay of the connection.

TS   Traffic shaping: Traffic shaping is a mechanism that alters the traffic characteristics of a stream of cells on a connection to achieve better network efficiency, while meeting the QoS objectives, or to ensure conformance at a subsequent interface. Traffic shaping must maintain cell sequence integrity on a connection. Shaping modifies traffic characteristics of a cell flow with the consequence of increasing the mean cell transfer delay.

TS   Transport stream: One of two types of streams produced by the MPEG-2 systems layer. The transport stream consists of 188-byte packets and can contain multiple programs.

TSAP   Transport service access point

TTCN   Tree and tabular combined notation: The internationally standardized test script notation for specifying abstract test suites. TTCN provides a notation that is independent of test methods, layers, and protocol.

UBR   Unspecified bit rate: UBR is an ATM service category that does not specify traffic-related service guarantees. Specifically, UBR does not include the notion of a per-connection negotiated bandwidth. No numerical commitments are made with respect to the cell loss ratio experienced by a UBR connection or as to the cell transfer delay experienced by cells on the connection.

UDP User datagram protocol: This protocol is part of the TCP/IP protocol suite and provides a means for applications to access the connectionless features of IP. UDP operates at layer 4 of the OSI reference model and provides for the exchange of datagrams without acknowledgments or guaranteed delivery.

UME UNI management entity: The software residing in the ATM devices at each end of the UNI circuit that implements the management interface to the ATM network.

Unassigned cells A cell identified by a standardized virtual path identifier (VPI) and virtual channel identifier (VCI) value, which has been generated and does not carry information from an application using the ATM layer service.

UNI User-network interface: An interface point between ATM end users and a private ATM switch, or between a private ATM switch and the public carrier ATM network; defined by physical and protocol specifications per ATM Forum UNI documents. The standard adopted by the ATM Forum to define connections between users or end stations and a local switch.

Unicasting The transmit operation of a single PDU by a source interface where the PDU reaches a single destination.

UPC Usage parameter control: Usage parameter control is defined as the set of actions taken by the network to monitor and control traffic in terms of traffic offered and validity of the ATM connection at the end-system access. Its main purpose is to protect network resources from malicious as well as unintentional misbehavior, which can affect the QoS of other already established connections, by detecting violations of negotiated parameters and taking appropriate actions.

Uplink Represents the connectivity from a border node to an upnode.

Upnode The node that represents a border node's outside neighbor in the common peer group. The upnode must be a neighboring peer of one of the border node's ancestors.

Usage parameter control (UPC) Prevents congestion by not admitting excess traffic onto the network when all resources are in use. UPC changes the CLP bit of cells that exceed traffic parameters so they are dropped.

User network interface (UNI)   The protocol adopted by the ATM Forum to define connections between ATM user (end station) and ATM network (switch). The UNI specifies the complete range of ATM traffic characteristics including cell structure, addressing, signaling, adaptation layers, and traffic management.

UT   Upper tester: The representation in ISO/IEC 9646 of the means of providing, during test execution, control and observation of the upper service boundary of the IUT, as defined by the chosen abstract test method.

UTOPIA   Universal test and operations interface for ATM: Refers to an electrical interface between the TC and PMD sublayers of the PHY layer.

UTP   Unshielded twisted pair: A cable having one or more twisted pairs, but with no shield per pair.

UTP   Unshielded twisted pair cable

Variable bit rate (VBR)   Information that can be represented digitally by groups of bits (as opposed to streams) is characterized by a variable bit rate. Most data applications generate VBR traffic, which can tolerate delays and fluctuating throughput.

VBR   Variable bit rate: An ATM Forum defined service category that supports variable bit rate data traffic with average and peak traffic parameters.

VC   virtual channel: A communications channel that provides for the sequential unidirectional transport of ATM cells.

VCC   Virtual channel connection: A concatenation of VCLs that extends between the points where the ATM service users access the ATM layer. The points at which the ATM cell payload is passed to, or received from, the users of the ATM layer (i.e., a higher layer or ATM-entity) for processing signify the endpoints of a VCC. VCCs are unidirectional.

VCI   Virtual channel identifier: A unique numerical tag as defined by a 16-bit field in the ATM cell header that identifies a virtual channel, over which the cell is to travel.

VCI   Virtual circuit identifier

VCI Virtual connection identifier

VCL Virtual channel link: A means of unidirectional transport of ATM cells between the point where a VCI value is assigned and the point where that value is translated or removed.

VCO Voltage-controlled oscillator: An oscillator whose clock frequency is determined by the magnitude of the voltage presented at its input. The frequency changes when the voltage changes.

VD Virtual destination. Refer to VS/VD.

VF Variance factor: VF is a relative measure of cell rate margin normalized by the variance of the aggregate cell rate on the link.

Virtual channel identifier (VCI) The unique numerical tag used to identify every virtual channel across an ATM network, defined by a 16-bit field in the ATM cell header.

Virtual channel switch A network element that connects VCLs. It terminates VPCs and translates VCI values. It is directed by control plane functions and relays the cells of a VC.

Virtual channel A defined route between the endpoints in an ATM network that may traverse several virtual paths.

Virtual circuit (VC) A portion of a virtual path or a virtual channel that is used to establish a single virtual connection between two endpoints.

Virtual path identifier (VPI) An 8-bit field in the ATM cell header that indicates the virtual path over which a cell is to be routed. A virtual connection established using only the VPI is known as a virtual path connection (VPC).

Virtual path switch A network element that connects VPLs. It translates VPI (not VCI) values and is directed by control plane functions. It relays the cell of the VP.

Virtual path A group of virtual channels, which can support multiple virtual circuits.

VLAN   Virtual local area network: Workstations connected to an intelligent device that provides the capabilities to define LAN membership.

VLSI   Very large scale integration

VP   Virtual path: A unidirectional logical association or bundle of VCs.

VPC   Virtual path connection: A concatenation of VPLs between virtual path terminators (VPTs). VPCs are unidirectional.

VPCI/VCI   Virtual path connection identifier/virtual channel identifier

VPI   Virtual path identifier: An 8-bit field in the ATM cell header that indicates the virtual path over which the cell should be routed.

VPL   Virtual path link: A means of unidirectional transport of ATM cells between the point where a VPI value is assigned and the point where that value is translated or removed.

VPT   Virtual path terminator: A system that unbundles the VCs of a VP for independent processing of each VC.

VS   Virtual source. Refer to VS/VD.

VS   Virtual scheduling: Virtual scheduling is a method to determine the conformance of an arriving cell. The virtual scheduling algorithm updates a theoretical arrival time (TAT), which is the "nominal" arrival time of the cell assuming that the active source sends equally spaced cells. If the actual arrival time of a cell is not "too" early relative to the TAT, then the cell is conforming. Otherwise, the cell is nonconforming.

VS   Virtual source

VS/VD   Virtual source/virtual destination: An ABR connection may be divided into two or more separately controlled ABR segments. Each ABR control segment, except the first, is sourced by a virtual source. A virtual source implements the behavior of an ABR source endpoint. Backwards RM-cells received by a virtual source are removed from the connection. Each ABR control segment, except the last, is terminated by a virtual destination. A virtual destination assumes the behavior of an ABR destination endpoint. Forward RM-cells received by a virtual destination are turned around and not forwarded to the next segment of the connection.

WAN Wide area network: This is a network that spans a large geographic area relative to office and campus environment of LAN (local area network). WAN is characterized by having much greater transfer delays due to laws of physics.

WATM Wireless ATM: An initiative within the ATM Forum to develop a specification for transmitting ATM over wireless links. At the August 1996 meeting of the Forum, the WATM working group began to discuss the technical aspects of the service. The group's goal is to develop a scheme for transporting voice, fax, data, and video via wireless ATM links over satellite and microwave systems. Still under discussion are the frequencies and data rates. The group predicts the first release of the specification will be in late 1998.

XDF Xrm decrease factor: An ABR service parameter, XDF controls the decrease in ACR associated with Xrm. It is a power of two in range: [0, 1].

XNS Xerox Network Systems

Xrm An ABR service parameter, Xrm limits the number of forward RM-cells that may be sent in the absence of received backward RM-cells. The range is 0–255.

XTP Express transport protocol

16-CAP Carrierless amplitude/phase modulation with 16 constellation points. The modulation technique used in the 51.84-Mbps midrange physical layer specification for category 3 unshielded twisted pair (UTP-3).

64-CAP Carrierless amplitude/phase modulation with 64 constellation points.

About the Author

Ed Coover majored in history in college (B.A., M.A.) and then served as a Marine Corps communications officer in Vietnam. Interested in statistics and social measurement, he supported himself as a computer programmer while getting the third degree in history (Ph.D.). He has taught at the University of Minnesota (Saint Paul), Indiana University (Bloomington) and Warsaw University (Poland). Finding no future in the past, he went to work in the computer industry. Over 19 years this has included an inept computer time-sharing firm, a large and rapacious systems house, a network analysis boutique, and, most recently, a Federally Funded Research and Development Center (FFRDC). Meanwhile, his professional interests have increasingly focused on computer networks. He has contributed to *Datamation*, *Data Communications*, *LAN Magazine*, *Telephony*, and the Institute of Electrical and Electronics Engineers' (IEEE) *Communications Magazine* as well as editing tutorials on PBXs and SNA networks for the IEEE. For fun he jogs, sails, cross-country skis, travels to strange places, sees lots of films, reads novels, listens to opera, and talks to his dogs. He has two grown daughters, is still married to his first wife, and lives in Chevy Chase, Maryland.

Index

The Artech House Telecommunications Library

Vinton G. Cerf, Series Editor

UNIX Internetworking, Second edition, Uday O. Pabrai

Videoconferencing and Videotelephony: Technology and Standards, Richard Schaphorst

Wireless Access and the Local Telephone Network, George Calhoun

Wireless Communications in Developing Countries: Cellular and Satellite Systems, Rachael E. Schwartz

Wireless Communications for Intelligent Transportation Systems, Scott D. Elliot and Daniel J. Dailey

Wireless Data Networking, Nathan J. Muller

Wireless LAN Systems, A. Santamaría and F. J. López-Hernández

Wireless: The Revolution in Personal Telecommunications, Ira Brodsky

Writing Disaster Recovery Plans for Telecommunications Networks and LANs, Leo A. Wrobel

X Window System User's Guide, Uday O. Pabrai

For further information on these and other Artech House titles, contact:

Artech House
685 Canton Street
Norwood, MA 02062
617-769-9750
Fax: 617-769-6334
Telex: 951-659
email: artech@artech-house.com

Artech House
Portland House, Stag Place
London SW1E 5XA England
+44 (0) 171-973-8077
Fax: +44 (0) 171-630-0166
Telex: 951-659
email: artech-uk@artech-house.com

WWW: http://www.artech-house.com